T0075927

Mobilizing Science

Mobilizing Science

Movements, Participation, and the Remaking of Knowledge

Sabrina McCormick

TEMPLE UNIVERSITY PRESS
Philadelphia

Sabrina McCormick is Robert Wood Johnson Health and Society Scholar at the University of Pennsylvania and holds a joint appointment as Assistant Professor of Sociology and of Environmental Science and Policy at Michigan State University. She is the author of *No Family History: The Environmental Links to Breast Cancer.*

Temple University Press
1601 North Broad Street
Philadelphia PA 19122
www.temple.edu/tempress

Published 2009
Printed in the United States of America

♾ The paper used in this publication meets the requirements of the American National Standard for Information Sciences—Permanence of Paper for Printed Library Materials, ANSI Z39.48-1992

Library of Congress Cataloging-in-Publication Data

McCormick, Sabrina.
Mobilizing science : movements, participation, and the remaking of knowledge / Sabrina McCormick.
p. cm.
Includes bibliographical references and index.
ISBN 978-1-4399-0009-3 (hardcover : alk. paper) 1. Science and state–Citizen participation. 2. Technology and state–Citizen participation. 3. Science–Decision making–Citizen participation. 4. Technology–Decision making–Citizen participation. 5. Science and state–Citizen participation. 6. Technology and state–Citizen participation. I. Title.
Q125.M416 2009
303.48'3–dc22 2008048817

2 4 6 8 9 7 5 3 1

Contents

Introduction

Claudia's[1] graying teeth and limp hair meant she came from the farmlands. Periods of intense drought made food scarce and nutrition limited. Those farmlands, I had heard, bred some of the most radical protestors. Claudia was a perfect example. A righteous attitude complemented her stubborn nature. Taking no for an answer was never an option. She told me that she had not married because it would hold her back; macho Brazilian men and their ideas about a woman's place were not worth her time or energy. She had instead given her life to the church in a way I did not understand. Not a nun, exactly, but a dedicated worker in the path of justice. Her church actively supported the downtrodden, the marginalized, the many people pushed off their lands by stronger corporate interests. As one whose lands had been exploited, Claudia was indignant and, despite almost no education, brilliant.

Claudia was one of the first people I met who protested the construction of large hydroelectric dams in Brazil. Our first encounter was at a roundtable held by local professors to discuss several dams with the community. I had not been to the interior of Brazil before and was fascinated by its inhabitants, not because of their customs or style of dress but because of their understanding of the land and their community. For the most part, they had never entered a formal

[1] Claudia is a pseudonym used to protect the actual individual.

school but had instead grown up learning the land. They knew how to make nourishment rise from lackluster soil and to catch the largest fish so that the smaller ones would be left for next season. They intimately understood the ebb and flow of the river from which they drew enough water to plant crops year after year. And they knew what the proposed dams would do to their river, their land, and their livelihoods.

Claudia spoke eloquently and passionately at this first meeting. I watched the professors learn from her, and she from them. Despite her experience with protest, she did not fully understand reports the state had generated to argue that her land and people should be displaced. Meanwhile the professors had little knowledge of the crops produced in the area, where community members lived, or exactly how the proposed dams would shift the local economy. As they explained the report's findings, she countered with what she knew to be the truth. The professors took notes. Her input was added to their critique of the inaccurate environmental impact assessment they held in their hands. The community did not have a copy. It had not, in fact, been offered one by the state. The researchers seemed to be the brokers of information, offering state studies to the community and working with this group to generate a new study. They would, in turn, offer the new impact assessment to the local environmental agency, suggesting that the dam not be built.

Although I had seen the anti-dam movement occupy construction sites and effectively slow the progress of dam building, I knew that behind-the-scenes collaboration was a tactic critical to the movement. Activists invested great amounts of time and resources in working with the researchers to learn about the environmental impact assessment (EIA) format, make a new report showing all the missed information, and then share their perspectives at a public hearing. The hearing where this private deconstructing and reconstructing of the environmental impact assessment would become public was the only pathway for citizens to participate in the policymaking process. In addition, since the EIA was the main justification for the dam, it was an important terrain of contestation.

At the hearing, arguments would be laid out by the local community, the government agency, and the representatives of the corporation that was funding the project. But the community would not have formal decision-making power. And ultimately, this community, which might be displaced by one of multiple dams being planned in the area, would most likely be flooded. Its participation in the process would make no difference, despite the organizing, educating, and collaborating. This case is not singular in nature. It represents a dilemma that spans the globe—movements engaging in long, drawn-out processes of changing research in order to shape policy, but facing the possibility of complete inefficacy. Democratizing science in this way is a growing movement tactic that is necessary because research findings have become the basis of policy making and the corporate justification for new products and practices. But this tactic is often co-opted through superficial participatory mechanisms and overpowered by large transnational corporations. Therefore, movements face a dilemma: how to democratize science without getting sidelined.

The Outcome of Dual Processes

Distrust of science and technology has arisen at the same time that public consumption of them has skyrocketed. Here I refer to "science" as the institutions that fund research, experts who conduct experiments and generate findings, and technologies whose impact may filter down to the public. In the United States, this resulted in protest after World War II when conflict-based technological advances came into question (Egan 2007). Contestation of science has also developed in industrializing nations as certain forms and institutions of science have become globalized (Fairhead and Leach 2003). This protest is in part due to the way science is generated. Much research and many resultant technologies, such as methods of energy generation, nanotechnology, large-scale farming, genomics, and chemicals, have been developed with little input from the lay populations whose lives they affect (Heller 2002; Macnaghten, Kearnes, and Wynne 2005). The "iron triangle" of government, corporations, and research has locked citizens out of political decision making about these products (Egan 2007). This situation has been exacerbated by the "scientization" of politics, where experts play a critical role in policy making and laypeople are marginalized.

Many movements protest science and related technologies with a particular focus on the lack of democracy involved in their development. These movements aim to democratize science, both the institutions that insulate it and the actual production of new technical information. Democratizing science movements (DSMs) contest expert knowledge and critique research findings as biased and politically driven. They aim to legitimate lay knowledge in science processes, government policies, and public discourse. They also intend to change the value structure underlying science, drawing attention to the biased nature of research trajectories and ignored impacts of new technologies. These actions have unintended consequences for movements, such as shaping the success of activists, research trajectories, and political functioning. Despite the growing exchange between science and movements, there is little understanding of how activists can effectively democratize science, and even less assessment of the social structures that make them fail.

This book examines social contestation of large hydroelectric dams in Brazil and another seemingly dissimilar case, movement formation around environmental causes of breast cancer, in order to examine how movements democratize science. Research on social movements, ethnoscience versus citizen science, the role of science in policy, and debates over corporate/government relations has generally been hindered by the divide between developed and developing worlds, ignoring commonalities between them. Democratizing science movements are sprouting everywhere. They demonstrate the interplay between "global science" and civil society (Fairhead and Leach 2003), as well as the similarities across DSMs around the world. A transnational, transtopical approach articulates the importance of contextual factors while highlighting the global nature of these struggles.

I trace the rise of democratizing science movements, explore why they are emerging and how they function, and examine under what conditions they are able to make change. I focus on these two cases, as well as examples of many others, to examine the before, during, and after of DSMs, including what they do when democratizing science fails. I argue that DSMs have emerged in response to government dependence on research and corporate control of it. The process of scientization has shaped the very functioning of movements, instigating movement participation in research projects and government institutions. Yet engaging in scientific production does not have clear outcomes. It may advance movement agendas or actually be a source of activist co-optation. Theorists of participatory research and of deliberative democracy argue that truly participatory measures are vastly difficult to create, and few institutions have actually been successful at it (Heller 2001a, 2001b; P. Evans 2002; Ryfe 2005). This is of particular concern in a "knowledge society" (Böhme and Stehr 1986) where science is essential to democracy but is not in itself democratic. This research asks if there is a way to alter entrenched epistemic philosophies and superficial governmental mechanisms through which social movements participate in research.

I also argue that these movements are only sporadically able to create such institutions, and that more often than not, participation is circumscribed. Several interlocking factors shape the success of these movements, including the scientific frameworks, methods, tools, and norms at play, the participatory parameters of government institutions, and the overwhelming power of transnational corporations that are mandated both to fund and to conduct science. In addition, movement tactics and approaches codetermine outcomes. DSMs have multiple goals and tactics, some of which are more effective than others. For instance, a DSM can more readily gain information about science and distribute it to constituents than challenge the policies of a federal government that are based on limited research funds. By detailing the successes and failures of these two movements, this research offers several resources to readers.

First, this framework offers an explanation of how the growing saliency of science and technology and resultant public disenfranchisement from them become a terrain of contestation. A social movement is a network of individuals with shared beliefs and agendas who work together to protest a set of institutions, practices, or social norms over a period of time. It arises when a group, generally defined by a certain identity, seeks to improve its welfare through changing policy, its public image, or its ability to achieve certain goals (Della Porta and Diani 2006). As science and technical information have become critical to improving health, welfare, and policy, movements have begun to use it in grievance formation. However, as activists gain knowledge of science, they often find that new research must be developed to support their hypotheses and goals. As a result, they engage in scientific production.

In order to understand why science and technical information have become fundamental to many movement struggles, we must look at how lay citizens are disconnected from research trajectories, how expert knowledge is inculcated in political institutions, and how corporate interests have developed power over it. Movements attempt to insert a new set of interests into research that represents marginalized groups. In other words, by focusing on DSMs, we can begin to see how movements intervene in processes where "science converts knowledge into power" and we can also examine "how power converts interests and power into science" (Harding 1998a, 51).

Like many movements that are not concerned with science, democratizing science movements attempt to address the privatization of conflict that limits the capacity of citizens to make change

(Schattschneider 1960). More specifically, they attempt to include the knowledge claims of nonexperts in order to overcome limitations of who is involved in a debate, or what Schattschneider calls "scope." Scope shapes movement outcomes. Schattschneider claims that privatization limits the scope of conflict; this process is similar to scientization. For DSMs, the idea of scope raises questions about the definition of an activist, especially how expert or nonexpert identities cross in untraditional ways. These issues are important in debates related to science or technology.

This leads to the second contribution of this text—a better understanding of how lay disagreements with official scientific codifications get introduced, publicized, and mediated. DSMs use multiple methods and have several different outcomes. Movements "reframe" research and scientific objectivity and then work to change science itself. Through the first tactic, activists work to change public opinion and understanding. Engaging in science is then used to alter research topics, advance findings, and change related policies. In both movement methods, activists work with scientists to gain access to information and credibility.

Finally, through these cases we can see how the intent of democratizing science shapes movements themselves. Social movement theories such as political process (McAdam 1982), framing (Gamson and Meyer 1996; Steinberg 1998; Benford and Snow 2000), and resource mobilization (McCarthy and Zald 1973) intersect with theories regarding the role of science in policy (Bimber and Guston 1995; Jasanoff 1990) to help elucidate how scientization engenders certain kinds of DSM tactics. On the other hand, the very process of engaging with scientists shapes movement outcomes. For example, the politicization of local knowledge encouraged by lay/expert collaborations can become a root of organizing and movement development. Also, focusing on science means marginalization of other tactics.

Using scientific contestation as a tactic may also influence to what extent a movement can achieve its agenda. In some instances, it may instead serve as a form of movement co-optation or lead to movement failure. I aim to differentiate when a movement will be co-opted and when it will be able to achieve its goals through democratizing science. Outcomes are partially determined by what type of deliberation and participation takes place. Because a great deal of research is government funded and because the science that informs policy is housed in political institutions, understanding the outcomes of social movements involves examining both participatory research and the structure of political institutions.

TABLE I.1 TYPOLOGY OF PARTICIPATION

Lay/expert form	Direction of knowledge transfer	Methods of knowledge-sharing	Main activities
1. Researcher educator	Top→ bottom	Expert language sharing	Researchers serve as educators of movement representatives or laypeople
2. Researcher activist	Top→ bottom	Construction of a new discourse outside collaborative space; instrumental legitimacy offered to support communicative claims	Researchers serve as movement leaders or political representatives
3. Citizen/science alliance	Bottom→ top	Discussion and deconstruction of official knowledge; countering of expert and lay claims	Construct new research about the impacts of dams or causes of breast cancer
4. Collaborative forum	Bidirectional	New, official codifications of knowledge	Laypeople and experts construct official documentation of dam projects or breast cancer research

In order to articulate different kinds of participation that span these multiple DSM approaches, I present a four-part typology that includes researcher educator, researcher activist, citizen/science alliance, and collaborative forum (Table I.1). These types are much broader than most participatory theories because they account for lay/activist engagement outside the realm of formal institutions and inside the space of movement organizing. They are surprisingly consistent across the two empirical cases at hand. These forums have the capacity to be new institutions of deliberative democracy that improve democratic decision making by involving citizens more thoroughly in many spheres, such as government and the economy (Warren 1996). Some of them are more effective than others, and all of them create different outcomes. This typology may be a resource for other researchers who investigate similar movement activities.

Theoretical Departures for This Book

This book provides several theoretical contributions. First, it articulates the direct interrelationship among social movements, science, and government institutions (Morello-Frosch et al. 2005). On the basis of the work of social movement theorists, it might appear that government institutions have been more porous to movements than scientific institutions have been. This appearance is driven by the smaller amount of attention devoted to movement/science interactions. More recent scholarship has drawn attention to their interrelationships and the influence that movements have had on science (Steven Epstein 1996; Egan 2007; Corburn 2005; Hess 2007; Moore 2008). I aim to add to these recent studies and create a mesolevel theory for them.

Second, I intend to clarify the forms of participatory research or political institutions that work best for movements across national context. In order to do so, I connect literatures regarding community-based participatory research (CBPR), participatory action research (PAR), feminist epistemology, and deliberative democracy. There has been a long-standing divide between these bodies of literature that can be overcome only by a cross-national comparison that brings these diverse theories together. Past research has avoided this bridging because such cross-national comparisons generally face serious methodological difficulties. Theoretically, however, they inform one another in important ways. Feminist epistemological theory is a linchpin of making a connection between CBPR and PAR; it shows that the knowledge possessed within oppositional consciousness crosses all kinds of boundaries, including race, nationality, gender, and subject of contestation (Harding 1998b). Combining theory on participatory research with research on deliberative democracy is essential because of the scientization of government institutions. Political dependence on science means that movements attempt to develop new or use existing government institutions to engage in research and policy making. In a knowledge society, we should not make a distinction between participatory science and participatory government. Science and government are too intimately interwoven for that. This book advances these bodies of literature by acknowledging the social context from which participatory projects emerge. Theorists interested in deliberative political institutions or research processes often forget to account for the social forces, like movements, that are pushing forward the creation of democracy.

The gap between the CBPR and deliberative democracy literatures is driven by the underlying assumption that laypeople, as so named by the first group, and citizens, as the second group of scholars refers to them, are subjects of participation or deliberation rather than actors who create it. The question has more often been how lay citizens function within these structures than how they instigate and shape them. This is where social movement theorists are helpful. They address the question of what movements can do. But they generally do not deal with questions about participation studied by the first two groups. In addition, few of these previously mentioned theorists address the actual process of scientific development or science as an institution.

Why Look at These Two Cases?

I use two case studies to fill out the details of the DSM framework: the environmental breast cancer movement (EBCM) in the United States and the anti-dam movement (ADM) in Brazil. The EBCM is a subset of the larger breast cancer movement that focuses specifically on potential environmental causes of the illness. It arose in three areas in the United States—the San Francisco Bay Area, Massachusetts, and Long Island, New York—that have some of the highest rates of breast cancer in the world. Unlike the broader breast cancer movement that focuses on raising funds to find a cure, the EBCM directs attention to prevention and causation. The anti-dam movement in Brazil is the only national-level movement of its kind. It protests undemocratic government practices that result in displacement and little compensation for poor, rural people. The movement has changed the conception of dams from sustainable to problematic and has simultaneously altered related national and international policy.

From a distance, these movements look entirely different, but they have a similar focus on democratizing knowledge. Although they are in very different countries, contest different topics, and are composed of very different populations, both movements find it necessary to contest expert knowledge and construct new research to advance movement goals. Their very differences highlight the similar importance of this movement method across contexts. I began studying the EBCM in order to understand how a small movement of women could create a completely new and controversial paradigm of illness causation, be successful enough to have their hypothesis tested with federal research funding, and gain enough attention to be reported on by national news sources (Kolata 2002). This question was the topic of my master's thesis at Brown University, for which I conducted research from 1999 to 2001 (McCormick 2001). I traveled to the three main locations where the movement had formed. I also spent the summer of 2001 working at the only research organization dedicated solely to studying women's environmental health, with a particular focus on breast cancer, Silent Spring Institute in Massachusetts. For more than seven years, I conducted interviews with social movement activists in local and national breast cancer organizations, government officials, and scientists who had collaborated with the movement. I also spent time with movement activists observing their activities and attending events across the country.

After studying the EBCM for some time, I began to come to conclusions about why a movement predominantly composed of highly educated, white, privileged women would challenge and manipulate, even engage in, science to achieve their agenda. They had access to political resources that funded the studies they wanted. They were generally not very radical, which meant that government funders were not threatened. Their education facilitated their understanding of scientific complexities used in studies. The movement worked in a societal context where expert knowledge had become so dominant that the only way to contest illness causation was to engage in science. In other words, they had advantages that gave them access to science, and they lived in a context that necessitated its use. This type of phenomenon, I assumed, would be isolated to a Western, democratized setting where movement actors had similar characteristics.

But as I finished my research, I began to wonder whether this assumption was accurate. In order to test it, I decided to compare the EBCM to a movement that was as different as possible on most counts—political context, activist identity, and subject being contested. I wanted to examine how different movement organizations could develop a similar focus on contesting and shaping knowledge and using it to make discursive and material change.

As I began my research, I realized that similar movements exist in places where science is much less institutionalized and normalized in everyday culture, where there is little participatory democracy, and where movement groups are almost completely marginalized. Researchers have reported on them in India, Africa, Eastern Europe, and Latin America. The variety of topics that these movements contest is vast, although many of them are environmentally related, like the topic of the EBCM. I selected the anti-dam movement in Brazil for comparison. It was different from the EBCM in almost every way but the one I was interested in understanding—contestation of science and democratization of knowledge. By focusing on this similarity, I hoped to understand why this focus persisted across contexts. I spent four years traveling to Brazil (2002–2006), spending two to four months every year crossing the length and breadth of the country. I began in the south, where the movement originally arose, interviewing activists, researchers, and government officials. Each group connected me with the next and led me to another region linked to activist networks. Eventually, I was able to meet anti-dam activists and their expert counterparts in all five regions of the country. Most of the findings in this book relate to that period of time, although they have been updated through 2007.

Like the EBCM, the anti-dam movement had allied with scientists to contest the expert knowledge used to justify dam planning. These lay/expert collaborations were taking place all over the country, creating an entirely new discourse about dam building and even stopping the construction of dams (Zhouri, Laschekski, and Barros Pereira 2005; Rothman and Oliver 1999). With the assistance of these laypeople, experts wrote books about how dams were highly problematic and unnecessary (Bermann 2002). As my interviewees described in depth, even government officials had begun to adopt this new discourse. While largely unconscious of the movement itself, the public had begun to believe that there were other, more effective options for generating energy (Goldenberg 2002). And in contrast to the case of the EBCM, the people most affected by dams were largely rural workers who lived in a country where science received only peripheral attention (Araujo 1990).

The ADM had long been characterized as a poor people's movement shaped by environmental conflict. The grassroots, protest-based nature of activism makes this apparent. Less obvious, however, is the role that science could possibly play in such struggles where livelihoods and ancestral histories are at stake, and where human rights are being violated. Most research on the role of science in movements has not addressed such a movement, yet conflicts are often mediated by the state through the arena of science, which is meant to be the arbiter of disputes. Violent conflict may continue between the state or corporations and activists at the same time the terms of development are negotiated through science.

Despite their vast differences, several factors drive the similarity of these two movements and many other democratizing science movements. The first is the expert basis of public perception and policy making. In these two cases, the expert control of the "war on cancer" (Sporn 1996) and the building of what Jawaharlal Nehru called temples of progress make the ADM and the EBCM more likely bedfellows than they might appear. For example, breast cancer activists point to unregulated chemicals in our environment that may affect human health (Breast Cancer Fund 2007). Whether or not each chemical is licensed supposedly depends on scientific measurement of its toxicity, although in practice most chemicals are not tested (Vogel 2004). Energy policy is also allegedly based on expert impact assessments (Fearnside 2006), with little consideration for the environmental, health, and social issues raised by displaced people.

The strength of political and economic interests that shape these bodies of expert knowledge and use it for their own purposes is an additional similarity between these two cases, as well as many others. Consequently, the EBCM and the ADM are driven to introduce new participatory structures to counter the influence of industry. In both countries, citizens are provided limited space to participate in the construction of environmental regulation. Industry has much greater influence on such policy than citizens (DeSombre 2000). Despite the existence of policies to regulate industry, these private interests have continually challenged passage and implementation of environmental regulation. Industry is in fact the main body that produces research necessary for a chemical or a dam to be approved by governmental agencies (Barrow and Conrad 2006). Independent institutes or consultants also do similar research, but often the researchers in those institutes are funded by industry (Yach and Bialous 2001).

Third, because of the vast array of internal movement differences, such as identity of movement actor and length of movement activity, I saw the similarity of focus on an environmental concern as helpful. The most common type of movements attempting to democratize knowledge is that addressing environmental concerns (Pickvance 1997). Both movements had combined environmental concerns with a social justice framework, although the resulting grievances were very different. Environmental movements have traditionally encompassed a large body of well-educated people, including scientists, who could use their expert assessments to counter other claims (Gottlieb 2005). This was necessary because of the increasing regularity with which environmental policy was based on science, research, and expert knowledge. As the environmental movement has grown and diversified to include many groups who are not led by experts, many local communities have come to argue that this bias toward expert knowledge removes them from a role in policy making and understanding of environmental degradation (Fischer 2000). Other fields fall prey to similar biases. Study of two movements that deal with environmental concerns and related expertise can aid assessment of other types of expertise in the future, be they legal, political, or of some other kind.

Finally, and most important, expert knowledge was central to both movements—their formation, tactics, and goals. Although movement actors had different educational backgrounds and each country invests different levels of resources in science, successes of both the ADM and the EBCM hinged on access to technical information and the ability of social movement activists to reformulate it. The following sections begin to describe the backgrounds of these cases and the major issues that they contest.

Breast Cancer: From Pink to Green

The influence of environmental groups in the United States has transformed public consciousness, turning trees into huggable symbols of life and toxics into a reflection of corporate control. Likewise, the breast cancer movement in the United States brought women's breasts onto the public agenda by literally placing them in the public eye (Casamayou 2001). Graphically depicted on billboards and magazine ads, the breast has become a symbol of motherhood, sexuality, and youthful vigor. Maybe at one time the idea of connecting these two concepts was an oddity, but the environmental breast cancer movement does this simply and strongly by arguing that women's bodies and breasts reflect the toxics in the environment as chemicals build up in breast tissues and result in cancer.

The environmental breast cancer movement has created a new public awareness of environmental causes of breast cancer, instigated massive government-funded research studies to examine environmental variables, and advanced a new conception of environmental health (McCormick, Brown, and Zavestoski 2003). The EBCM works toward four goals: (1) to broaden public awareness of potential environmental causes of breast cancer; (2) to increase research into environmental causes of breast cancer; (3) to create policy that could prevent environmental causes of breast cancer; and (4) to increase activist participation in research. It is focused on three main locales that have significantly higher incidences of breast cancer than the rest of the United States (Aschengrau et al. 1996, Robbins, Brescianni, and Kelsey 1997).

This activism began in Long Island and then spread to Massachusetts and finally to the San Francisco Bay Area (Klawiter 2000). Within each location, a number of organizations work locally and nationally. Some lean toward collaborating with groups nearby, while others form coalitions across the country. In Massachusetts, the three most important organizations are the Women's Community Cancer Project, the Alliance for a Healthy Tomorrow, and the Massachusetts Breast Cancer Coalition, the last of which founded Silent

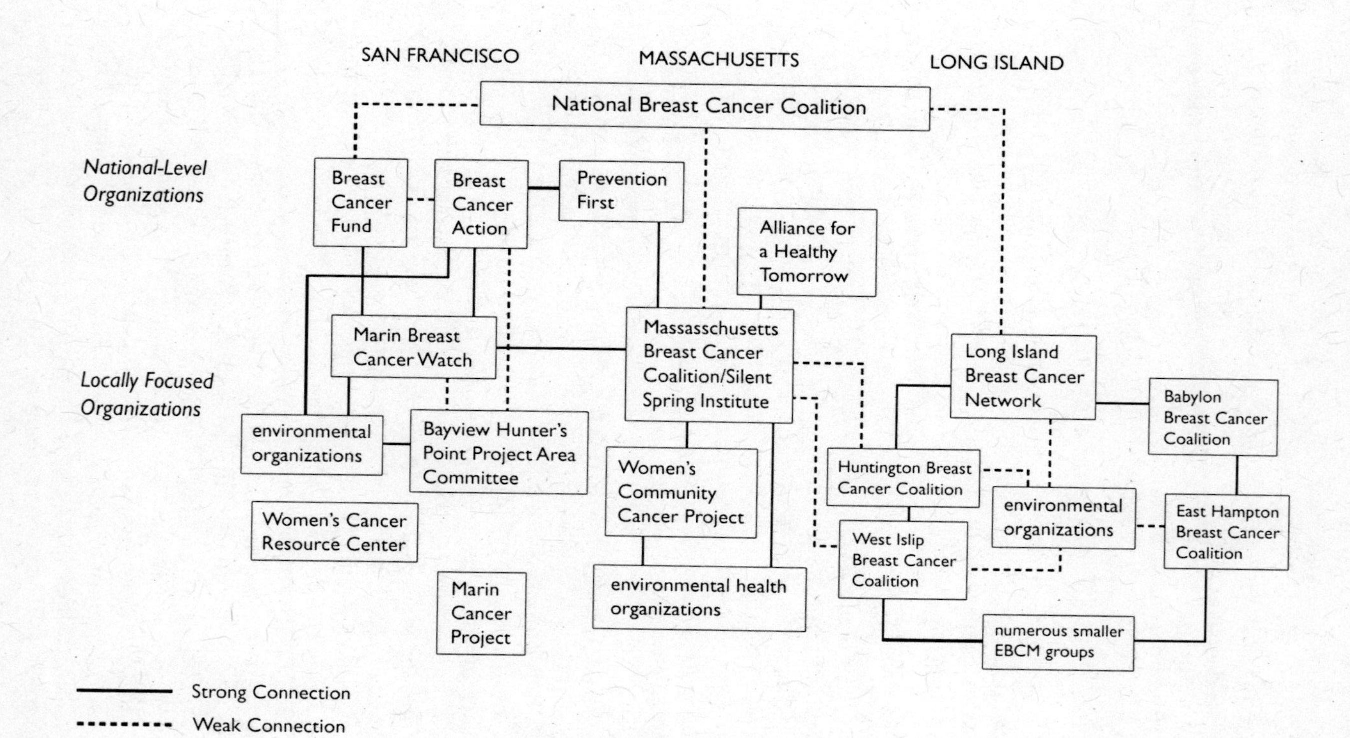

Figure I.1 EBCM Movement Organizational Chart

Spring Institute, which is devoted to studying environmental causes of breast cancer. On Long Island, a number of small service-based organizations have also engaged in science, such as Huntington Breast Cancer Coalition (whose name recently changed to Prevention Is the Cure), One in Nine, Babylon Breast Cancer Coalition, Breast Cancer Help, and West Islip Breast Cancer Coalition. These groups organized under one umbrella, the Long Island Breast Cancer Network, in order to collaborate with scientists on the Long Island Breast Cancer Study Project, the first major environmental breast cancer research project. In California, Breast Cancer Action, the Breast Cancer Fund, the Marin Breast Cancer Watch, and the Bayview Hunters Point Community Advocates are the main organizations that work both nationally and locally. Although local culture variations make the locales in which activism has occurred somewhat different (Klawiter 2001), the collectivity of knowledge and the overall movement ideology offer sufficient similarities to justify viewing this as a coherent national movement. These organizations also often engage in nationwide campaigns that unite their efforts. Figure I.1 visually depicts the level on which these organizations generally work, as well as their connections to one another.

Although the EBCM has focused mainly on three locales with high rates of breast cancer, it has also responded more generally to a world of cancer research that is predominantly focused on treatment, detection, and cure. This trajectory of research is a part of the larger "war on cancer" that Richard Nixon began in 1972 (Sporn 1996). This call was initiated by public concern about rising rates of cancers, of which breast cancer is one of the most common. Today in the United States, a woman is diagnosed with breast cancer every three minutes, and another woman will die of breast cancer every eleven minutes (N. Evans 2006). There were 178,480 new cases of invasive breast cancer and 62,030 cases of in situ breast cancer diagnosed in 2007, and 40,460 women died (American Cancer Society 2007).

Large Hydroelectric Dams: From Local to Global

Anti-dam groups began in the south of Brazil in the 1970s and grew into larger movements and organizations in the 1980s (Khagram 2004). The south faced construction first during the military dictatorship, when arbitrarily displaced small landowners were generally not powerful enough to contest construction. However, the opening of political opportunities as the Brazilian government transitioned to democracy between 1979 and 1983 stimulated movement activity (Rothman 1993). As more dams were built and resisted by local communities, these groups eventually began to exchange information and coalesce.

The anti-dam movement in Brazil has three main goals: (1) to alter public understanding of dams and their alternatives, (2) to increase democratic participation in the planning of energy generation, indemnification, and resettlement, and (3) to introduce a new energy model. The movement must use several tactics, including organizing, demonstrating legitimacy of its claims, generally through research projects, and protesting and demonstrating publicly. Strong grassroots groups and nongovernmental organizations (NGOs) work on multiple political and social fronts to achieve this mission.

National, regional, and local organizations work in conjunction with one another. Although local protest had formed in response to the construction of Itaipu in the south (Bruneau 1986), the first formal anti-dam organization was the Regional Commission of People Affected by Dams (Comissão Regional de Atingidos por Barragens, CRAB). It developed into what is now called the Movimento dos Atingidos por Barragens, or the Movement of Dam-Affected People (MAB). Additional important grassroots anti-dam organizations include the Movement for Development of the Trans-Amazon and Xingu River (Movimento Pelo Desenvolvimento da Transamazonica e Xingu, MDTX), based on the northern region, and Forum Carajas, based in the northeast but functioning across the country.

While MAB's organizational strategy has followed in the footsteps of the Workers' Party and the Movement of Landless People by creating small subunits in local areas where dams are being built (Wright and Wolford 2003), MAB and these other organizations also work with small, independent organizations, such as the Association of People Affected by Dam Projects in Gatos and Sucos (AMOGAS) in the interior of the northeastern region of the country, and groups of fishermen, indigenous people, and *quilombos* (isolated communities formed by escaped slaves). Finally, a series of international NGOs work across the country. They include the International Rivers Network (IRN), the Energy Working Group (Grupo de Trabalha de Energia, GT Energia), Living Rivers (Vidágua), SOS Matatlântica, and Greenpeace. Figure I.2 shows an organizational picture of the movement presented here for the purposes of understanding networks and relationships found in this research. Weak ties represent infrequent communication and collaboration and lack of real membership over-

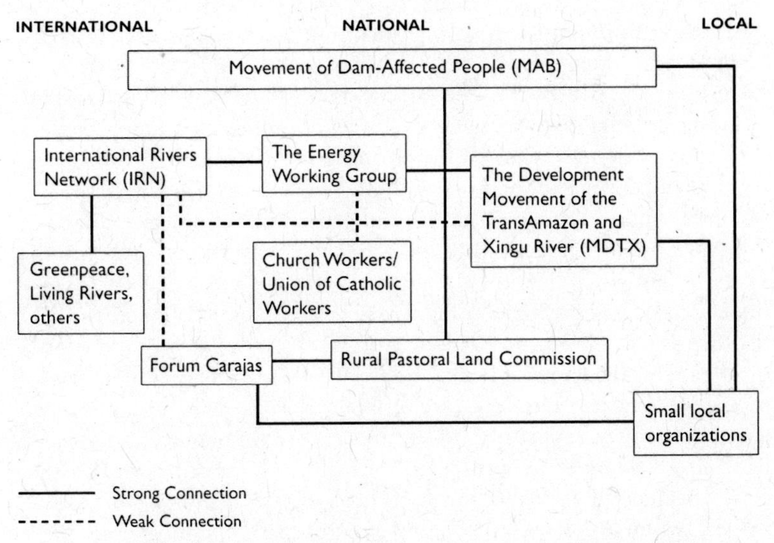

Figure I.2 Organizational Chart of the Anti-Dam Movement

laps, whereas strong ties signify sharing of substantial numbers of members, collaboration on projects, and regular interaction between group leaders.

The anti-dam movement has responded to the huge impact of hydroelectric dams in Brazil. Brazil has the third-largest installed hydroelectric capacity in the world. It has the greatest potential of any nation; larger than some entire continents (Santos, Andrade, and Wright 1990). Today, Brazil has more than six hundred dams (Brown and McCully 1997). Large dams are internationally defined as being higher than fifteen meters (LaRovere and Mendes 2000. They take ten to twenty years to construct and are costly, as is exemplified by the Itaipu Dam shared by Brazil and Paraguay, which was built in eighteen years and cost nearly $18.3 billion (Luxner 1991). Small dams are constructed much more frequently than large dams and have a related set of problems (Zhouri 2003).

Although a few large dams such as Tucuruí and Balbina were built in the north in the 1970s and 1980s, most dams in Brazil are in the southern and southeastern regions, where rolling hills are dotted by small farms. A long legacy of energy-intensive mining in the southeast was responsible for the earlier construction of dams, but many new ones are driven by the massive influx of industry. The majority of hydrological resources in the south are already used, and because of the construction of Sobradinho and Itaparica, the northeast has only a minimal amount of remaining river resources (Agencia Nacional de Aguas [ANA] 2003). Therefore, the three other regions have been the site of the newest construction. The first inroads of building began a few years ago in the center and west and are beginning to gain strong social and environmental attention.

The northern region, more than any other in Brazil, has the most potential and is characterized by the greatest risk. The presence of the Amazonian forest in the north is causing growing contention. Government documents estimate between twelve and sixty dams under consideration for construction. These plans were once part of the Avança Brasil project meant to initiate multiple megadevelopment projects in the area (Laurance et al. 2001) and are now considered necessary for national development. Resolution of the pending energy crisis was a part of President Luiz Inacio Lula da Silva's platform in 2002, and the construction of dams in the Amazon has recently been emphasized as a way to meet energy demand. The Amazonian region holds more hydrological potential than most countries in the world (McCully 2001). Yet because it is also the home to what many have called the lungs of the earth (Gore 2006), referring to the oxygen-producing capacity of the forest, it is consequently shielded by the international norm of environmental protection.

Vast Differences, Important Similarities

Because the EBCM and the ADM are vastly different in their internal functioning and contextual circumstance, concepts that are applicable to both are likely to be relevant to cases in many other places. The subjects of contestation are vastly different and have shaped the identity of movement actors. Because breast cancer largely affects women, women constitute the majority of the EBCM. Since dams largely displace poor, rural people, they make up the largest number of anti-dam activists. The structure and role of the state and each country's place in the global world system are also very different from one another.

Although they diverge on many counts, the ADM and the EBCM are somewhat similar in tactical approach. Like many if not most contemporary social movements, they both challenge the state, target the role of industry, engage in public protest, and organize or educate the public. These similarities may be caused by a long history of active social movements in both Brazil and the United States (Jaquette 1994; Treece 2000; Langer and Muñoz 2003). In Brazil this is signified by the critical role that social movements played in initiating democracy and the important role they continue to play in making democracy work (Cardoso 1992). In both Brazil and the United States, there are many social movement organizations that address a breadth of topics and provided lineage for the two movements studied here.

In the United States, a powerful feminist movement preceded the women's health movement (Altman 1996), which was followed by the general breast cancer movement, the environmental movement, and finally the environmental breast cancer movement. In Brazil, a strong movement of landless people and workers, then an indigenous people's movement, and an environmental movement have gained strength. The church has also traditionally facilitated movement advances as well (Eckstein 1989). All these movements assisted in the formation of the anti-dam movement.

Both the EBCM and the ADM make similar attempts to bring local perspectives into the debates leveraged against the state or industry and for the public. These movements use a particular similar mechanism—lay/expert collaboration—to change policies, make industry more accountable, create new frameworks of contestation, and educate community members about the topics they contest. They both aim to create participatory processes by challenging the shortcomings of representative democracy and offering citizens more participation in the governing process. Although the specific type of knowledge they contest varies, both movements attempt to increase valuation of lay knowledge and to penetrate insulated arenas of decision making. Studying the tactics through which these movements democratize knowledge reveals similarities in mechanisms of contention (Giugni, McAdam, and Tilly 1999).

Plan of This Book

The following chapters explore how democratizing science movements become successful or are co-opted. In the first chapter, I examine previous research on movements that engage in science and the gaps that research has left. In response, I describe an overall theoretical framework of DSMs that includes reasons why democratizing science movements emerge, and how the very contexts that engender their development limit their success. The following two chapters introduce the two case studies in depth. Chapter 2 tells the story of the founding of the environmental breast cancer movement, the role that scientists played in that process, and the scientific basis upon which it formed its framework, while Chapter 3 relates those details

TABLE I.2 MOVEMENT SIMILARITIES AND DIFFERENCES

	EBCM	Anti-dam movement	Common frame
Main Purposes	~Educate the public about environmental causes of breast cancer ~Promote research of these causes ~Push for participation of women with breast cancer in this research ~Prevent these causes/precautionary principle	~Educate/organize the public to fight dam construction ~Propose an alternative model of energy production ~Push for democratic decision making ~Push for resettlement where necessary	~To create participatory processes that result in the reduction of industrial impacts on the environment and people
Connections Made	~Health and the environment	~Environment and social exploitation	~Must consider marginalized populations (e.g., women or poor, displaced people) in the construction of environmental policy
Role of Knowledge	~Women have embodied knowledge of environmental exposures ~Scientific research marginalizes this knowledge	~Local people know most about their environment ~Their traditional knowledge is destroyed by dams/development	~Value of embodied lay knowledge is critical
Forms of lay/expert collaborations	~Participatory research institution formed (local and national levels) ~Research projects funded by the state	~Participatory research institution formed (local, national and international levels) ~Research projects funded by the state and foundations	~New research and institutions founded around participatory processes
Perspective on Industry	~Not properly regulated; therefore, it causes public health problems	~Is exploiting people and the environment	~Must slow or change the methods of industry ~Need to democratize industrial production decision making to allow for more public accountability

for the anti-dam movement. Both chapters show how collaborating with scientists not only serves the purpose of changing science or policy but also contributes to movement development and sustenance. It is critical to initial movement formation, the politicization of local community consciousness, and the formation of a community around new grievances that lead to sustained movement activism.

Chapter 4 looks behind the science and movements in order to reveal how government institutions and corporate interests overlap in research trajectories that raise movement concerns and instigate challenges. The paradigmatic approach and intersecting interests of government and industry often instigate movement creation of new scientific frameworks. The EBCM developed a new environmental connection that politicized the illness and argued for new types of research and regulation. In the ADM, a developmental emphasis that prioritizes economic gain has guided dam development and planning. The movement proposes a sustainable paradigm that makes planning accountable to local communities and reduces environmental impacts. These movements create such critiques by learning complicated science and translating it to their constituents.

Chapter 5 examines the lay/expert collaborations that are often the root of both learning to reframe science for the public and changing scientific trajectories. The two movements represented in this research have developed multiple collaborations that change scientific tools, methods, and topics in order to address activist concerns. By examining how these collaborations have functioned and connecting them to theories of participatory research, I assess how these projects can take place most effectively for both scientists and activists. These collaborations also often result in another unrecognized way in which collaborating with scientists advances movements: movement reframing of issues. This issue is also discussed in Chapter 5.

These collaborations face major challenges in the realms of science and politics. Scientists are often reluctant to engage and have their credibility challenged. Others critique participatory research as lacking objectivity. In the process of participating, activists often face the possibility of being co-opted by superficial participatory mechanisms that allow their involvement but do not give them decision-making power. Chapter 6 presents cases of movement co-optation. In the northern region of Minas Gerais, Brazil, local communities engaged in the public hearing process to protest the possible construction of a number of industry-funded dams to little avail. The cases at hand demonstrate how powerful interests shape mechanisms of participation and consequently necessitate movement challenge. Theories of deliberative democracy articulate best why these participatory mechanisms are inadequate and how they can be advanced.

Chapter 7 looks at the most recent developments in each movement and raises questions about what these new, long-term struggles and projects mean for their future success. By looking at the twenty-year battle over Belo Monte and the new National Institute of Environmental Health Sciences Breast Cancer and the Environment Research Centers, I begin to explore what these movements need in order to succeed and what their achievements might look like in the future. In the final chapter, I come to some conclusions about why democratizing science movements have created new tactics to accommodate the importance of research in a "knowledge society," and how they might be more effective. This requires a nuancing of different types of achievements that these movements can make. Ultimately, I conclude that the partnership between scientific experts and the affected public holds great potential for democratic practice and social norms in many countries. Democracy is most deepened when barriers to disenfranchised populations are overcome and their perspectives are incorporated into political discourse. However, private and public interests more powerful than movements have found ways to sidestep change. Only through recognizing how superficial answers are given to concrete demands, and the role of science in that process, can democracy be deepened.

1 Democratizing Science Movements

Conditions for Success and Failure

This chapter describes what democratizing science movements attempt to do and why they may fail. Scholars have begun to report on these movements around the world and to theorize movement functioning. I build on their work to create a general typology of democratizing science movements. Although they may have very diverse goals, I argue that they have a common impetus—the process of scientization—that may stimulate them to arise and also cause them to fail (Voss 1996). Often, activists are calling for greater democracy of research and government, but success depends on whether activists can effectively participate in political and scientific institutions rather than be co-opted through superficial participation. Co-optation "becomes possible when a challenging group or social movement opposes the practices, initiatives or policies of more powerful social organizations or political institutions" and the result is "some mix of institutionalization, social control, cooptation, and policy changes" (Coy and Heeden 2005).

What has been learned about social movement co-optation through political participation can be applied to scientific participation as well. Lay citizens may engage in research without formal decision-making power or in situations where they are greatly outnumbered by experts. In those cases, their perspectives may not be represented in outcomes, although their participation is celebrated.

Making the Case for Democratizing Science Movements

Movements around the world challenge expert knowledge, critique the minutiae of new technologies, and reshape research. On the continents of Asia, Africa, and South and North America, these movements address a diverse set of issues. Many of these are focused on health or the environment—from AIDS (Steven Epstein 1996) to pollution regulation (Hsiao and Liu 2002)—but others expand beyond those realms to address a multitude of political and corporate practices. Activists in Latin America have used "participatory research" as a movement strategy (Flint 2003). In line with Paolo Freire's conceptualization of "pedagogy for liberation" (1970), politically disenfranchised communities are empowered through learning and sharing their knowledge. A variety of movements in Europe have created "science shops" where research collaborations address movement concerns of many types (Leydesdorff and Ward 2005).

These movements have been active for some time, possibly because of their relevance for diverse topics. Although there is little information to trace their emergence and rise, it is likely that they largely emerged with the second wave of environmentalism that was accompanied by engagement with research discovering environmental degradation. Rachel Carson, the initiator of that movement, was possibly one of the first to call for them and be a part of them. Barry Commoner, another instigator of modern environmentalism, also championed democratizing science by helping lay citizens gain access to information about health effects of nuclear fallout (Egan 2007; McCormick 2008). In order to distribute more understandable information, Commoner and his group, the Committee for Nuclear Information, began to work with a group of women who had begun to be concerned about the health effects of nuclear fallout (Moore 2008). Soon after, environmental justice advocates arose and pointed to the importance of equal participation in research and decision making about environmental exposures.

Experts have been involved in environmental justice movements' attempts to address inequalities of environmental illness (Bullard 1990; Corburn 2005). Environmental justice movements direct attention to asthma or other chronic conditions of people of color disproportionately exposed to polluting sources. In order to demonstrate the connection between these sources and illness outcomes, they often engage in collecting data and forming scientific hypotheses. For example, "bucket brigades" gather air samples next to oil refiners and other polluters and then use those data to make arguments about exposure levels (Allen 2003).

Environmental justice activists have also attacked the norms that are embodied in scientific language, tools, and practice. Scientific norms are institutionalized in the same ways as norms about gender and race. In fact, these norms sometimes intersect. Asthma movements in the United States exemplify how movements contest science to affect racism (Brown et al. 2003). Organizations have formed in the Roxbury area of Boston and in Harlem in New York City in response to extraordinarily high asthma rates of inner-city children. These citizens protest the hazardous exposures in their community that they believe to be causing asthma. They have developed relationships with researchers at Harvard and Columbia to study environmental exposures in the community. They argue that predominant scientific paradigms guiding asthma research are racist and blame the victim; consequently, scientific findings do not address their suspicions and reinforce institutional arrangements that marginalize residents. In these ways, environmental justice movements very much exemplify the principles of DSMs.

Movements more specifically focused on health have engaged in scientific research because of the need to provide services or medications to constituents or to change perception of illness. In the case of AIDS activism, this agenda was driven by the need to eradicate prejudices embedded in scientific paradigms. Steven Epstein's (1996) work describing the AIDS movement provides the background for the environmental breast cancer movement's development of techniques to democratize research production and also theorizes movement/science relationships in a way that helps articulate the DSM framework. AIDS activists were some of the first to gain specific roles in government reviewing and granting of health research. They addressed the inadequacy and homophobic nature of certain types of scientific approaches and gained access to agencies and institutions in order to inform and shape research. AIDS activists were in a unique situation because of the emergence of a new, highly controversial disease. This opened the credibility arena to create, in Epstein's words, "multiplication of the successful pathways to the establishment of credibility and diversification of the personnel beyond the highly credentialed" (Epstein 1996, 3). EBCM activists followed this example. Activist interviewees even articulated that they learned participatory research tactics and the need to make demands on scientists from the AIDS movement. However, like most illness sufferers, they encounter an illness with an entrenched etiology. Therefore, they have faced greater barriers to transformation and also potentially have more lessons to offer about multiple illnesses.

Beyond Health and Environment

Many movements concerned about science work outside health and environment frameworks. These movements broaden the conception of what DSMs contest. Some DSMs have used critiques of research to supplement their overall agenda, others have made it their sole focus. The Kerala Shastriya Sahitya Parishad (KSSP) in India has changed scientific functioning or made it more legible to lay citizens. The KSSP is an organization that developed into a movement by popularizing scientific writings and providing scientific education for the masses in order to create more deliberative development practice. Zachariah and Sooryamoorthy (1994) demonstrated how science was a resource for activists attempting to create social action against development. Gaining access to technical knowledge also helped develop more participatory development practices. "Science shops" in Europe are different from the KSSP in that their main focus is on changing research itself. Initiated in the 1970s, they spread across Europe in order to connect university researchers with citizen interests or needs (Leydesdorff and Ward 2005). Twenty-one countries have hosted participatory projects, some of which are even funded by the European Union. Although these organizations have not traditionally been described as a part of social movement activity, they do represent sustained action where science is central to achieving goals.

Contesting science and the political institutions that use it is often a way to protest corporate power and practices. David Hess's (2005) research on social movement engagement with science includes movements dealing with transportation, organic foods, and energy. Technology- and product-oriented movements (TPMs) form coalitions with the private sector in an attempt to advance certain kinds of alternative technology. Hess highlights the importance of technology, not just research or science, to social movements. However, these movements are in some ways different from democratizing science movements because "their principal means of social change is the development of new or alternative forms of material culture" (2005, 517). On occasion, they change corporate products that are absorbed by the private sector and then transformed into something more profitable to industry. This is a type of co-optation faced by these activists.

Democratizing science movements also use the contestation of expert knowledge to make corporations more accountable to their concerns. For example, Heller (2002) demonstrated how farmer activists in Europe concerned about genetically modified organisms (GMOs) transformed the expert-generated scientific risk framing of GMOs to one instead focused on farmers' knowledge and the importance of food culture. Scientific framing was a proxy for the battles between two sets of actors—large-scale corporate brokers profiting from GMO sales and local producers. In this example and many others, science is a nexus between corporate, political, and activist interests. The manipulation of and contest over expert knowledge and technology are pathways through which these multiple interests intersect.

These authors clarify important points about the novelty of movements dealing with science. They show that science is a proxy for corporate interests, that citizens and scientists work together to generate new research topics, that often these collaborations change the very basic functioning of science, that marginalized actors raise these protests in an effort to change public, scientific, and government paradigms, that this is true in many national contexts, and that concerns about science represent issues of race and gender equality. Pulled together, these disparate pieces paint a picture of science reflecting the powerful and powerless. As such, it is a terrain of contestation that mediates differences between multiple groups in civil society, government, and corporate interests.

Defining Democratizing Science Movements

The rest of this chapter offers a framework of democratizing science movements that brings these bodies of research together. I outline their characteristics, including their subjects of contestation, goals, tactics, and strategies. The diverse constituents of these movements challenge how expert knowledge is produced, democratize knowledge production, and reframe expert knowledge. In the process, activists attempt to gain new credibility through their access to technical information, scientific language, and experts. For example, this activism can manifest itself in social movement leaders learning to write technically proficient newspaper editorials in response to the release of scientific studies or learning the technicalities of research methods so they can advise experts.

Democratizing knowledge—legitimizing lay knowledge in policy making, public discourse, and research or scientific discourse, usually

TABLE 1.1 DEMOCRATIZING SCIENCE MOVEMENT CHARACTERISTICS

Characteristic	Functions
Subject of contestation	Specific study, scientific paradigms or methods, form of scientific institution
Goals	To change the methods, content, institutions, or political impacts of science
Tactics	Participation in science, public protest and education, state challenge, identity-based organizing
Collaborative structures	Lay/expert collaborations
Stimulus for forming	Corporate control of science; lack of participation in political decision making

through lay/expert collaborations—is often an important part of these efforts. Activists can insert themselves into scientific spaces in order to shape research development, as in the increasing lay participation in peer review for the National Institutes of Health. They can also make broad claims about the needs for new types of science and knowledge based on their personal experience (e.g., suffering from a particular illness or being personally unprotected by a lack of governmental regulation).

The DSM framework can be used to describe a movement that is focused solely on science or to articulate how part of a particular movement functions (Table 1.1). While a movement may have a goal that is unrelated to science, it may use the tactics and have some of the goals of a DSM. Other movements may form with the explicit purpose of challenging and manipulating science. In other words, there is a range of types of DSMs, as reflected in Figure 1.1. These latter movements are less common than the former, but the fact that both exist indicates the relevance of the DSM framework as a transnational, transtopical phenomenon.

Social movements around the world have long challenged the state (Giugni 1998; Foweraker 1995), attempted to shift social norms

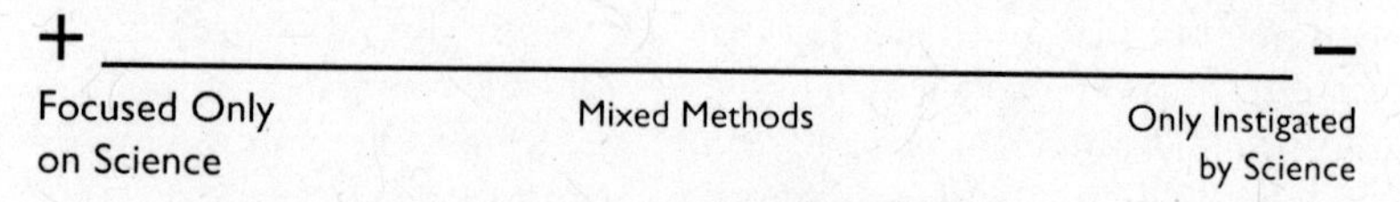

Figure 1.1 Range of DSMs

(Taylor 1989), or tried to improve social conditions (Castells 2004). DSMs use the contestation of experts, research trajectories, standards, tools and ethics of research, and technologies to achieve these same goals. In this way, they are both similar to and distinct from other movements.

The Players—Experts, Scientists, and Lay Citizens

Research conducted by experts is now the basis of most policy making (Jasanoff 1990). It also shapes our everyday decisions by indicating what to consume and how to act (Miller 2004). Official studies inform how we look at the world and how we talk about it. Political and economic actors use research to support their vested interests and consequently use findings to shape policy making (Jasanoff 2006). Experts are the brokers of information critical to government institutions, corporations, private interests, and social movements. They represent these various social actors. In this sense, contestation is a vastly diverse field. A scientist can be defined as "a researcher who, with much work, 'detects' something specific within nature" (Biagioli 1999a, 15). Scientists possess a certain kind of authority, have a unique training and set of credentials, and control a body of supposedly objective and value-neutral findings that can be leveraged by other social actors. Scientists are, in part, distinct from other experts because of the world they engage in and the knowledge they produce. Credit is generally assigned to that person or group of people who "detect" new phenomena but can also be attributed to the "corporation that paid the producer for his or her labor or rights in those claims" (Biagioli 1999a, 16). Therefore, scientists can be conceived of as both independent knowledge producers and those paid by a certain institution or set of institutions.

Scientists are part of an "epistemic community" (Haas 1992), or a loosely bound set of "social circles" that correspond to one another (Collins 1999). They use a number of tools, such as instruments and labs, that shape their work. Many of these elements are standardized and "inhabit several intersecting social worlds and satisfy the informational requirements of each" (Star and Griesemer 1999, 509). They are "simultaneously concrete and abstract, specific and general, contentionalized and customized" (517). As a result, many types of scientists may use similar tools and standards, and social actors refer to these "boundary objects" in processes of contestation. All these characteristics set scientists apart from technocrats or bureaucrats who have similar social positions based on political expertise.

Scientists are distinct from government bureaucrats, and this makes engagement with and contestation of their work a unique challenge. Bell (1999) claims that rather than bureaucracy, knowledge and technology are the main structural elements of the postindustrial society, where a growing proportion of society is funded to create new knowledge. A knowledge class composed of scientists predominates and forms an educated elite that is "at sharp variance with the population as a whole" (232). He offers a restricted definition of this expert knowledge—an objectively known intellectual property that is an investment by society and is subject to market judgments. In this sense, the category of "experts" can represent those who are paid for their expertise and who possess a certain kind of training that provides them with the legitimacy of that expertise. The main concrete importance is that they are able to generate new bodies of data as well as effectively evaluate existing expertise. They generally use what Francis (2001) calls etic knowledge. He distinguishes between two types of knowledge, that generated by those who observe, or *etic,* and those who act, *emic.* This is possibly the most basic differentiation between experts and others.

Etic knowledge can be juxtaposed with the idea of "metis"—local, lay, or "embodied" knowledge that is said to be subjectively known, acquired through practical experience not necessarily paid for financially, and often actually marginalized in markets. For example, knowledge of the environment is often gained simply through living in a locale, is not codified in any set way, and is generally not considered in the evaluation of cost-benefit analysis. Metis "represents a wide array of practical skills and acquired intelligence in responding to a constantly changing natural and human environment" (Scott 1998, 313). The concept of "embodied" knowledge acknowledges the diverse and power-differentiated communities in which knowledge originates (Haraway 1988). Marginalized actors who are not legitimated in scientific discourse possess embodied knowledge. Lay knowledge is possessed by social movement representatives or affected people. It can be likened to Nelson's (1990) conception of knowledge formation as based in community membership and on public conceptions of evidence. The acknowledgment of and turn toward these knowledge producers does not mean that they are inherently more accurate. Rather, as Harding claims, "Cultures' different locations in heterogeneous nature expose them to different realities of nature . . . and . . . such exposure to local environments can be a valuable resource for advancing knowledge" (1998a, 64).

Therefore, the usage of local or lay knowledge is less about whose knowledge is more accurate and more geared toward the recognition that there are multiple bodies of knowledge and value systems tied to those set of understandings. Harding (1998b) makes a strong claim that expert and lay knowledges should not be dichotomized, because doing so repeats Eurocentric patterns of privileging some forms of knowledge and knowledge production over others. She argues that instead, many systems of knowledge production should similarly be classified as science.

On the basis of this conception, some experts ally themselves with movements, while others represent the interests that movements protest (Tesh 2000). Some have lent their expertise, built new scientific arguments, and garnered scarce state support in the form of research funds (Fischer 2000). These coalitions are, in some senses, the link between science and society, between the creators and users of technology, and the window through which a new ethic of scientific research is being coconstructed (Quigley 2006). These collaborations have legitimated the involvement of nonexperts in scientific construction and have ultimately influenced official expert discourse. Through democratizing science, normative constraints on what is considered official knowledge are opened to make expert science more accountable to affected populations (McCormick et al. 2004; Cornwall 1995). A new body of research also emerges that often counters corporate- or state-funded science.

Fields, Tools, and Institutions of Science

Science defines itself with a sense of cohesion that disguises the diversity of fields and disciplines that it encompasses (Biagioli 1999a). In the language and terms of the international institutions that organize or fund science, like the United Nations Educational, Scientific, and Cultural Organization (UNESCO) or the National Science Foundation (NSF), the field of science includes scientists, social scientists, and engineers in a variety of fields who generate publications and patents, and who are distinct from premodern or indigenous knowledge (Drori et al. 2003). I follow in Hess's (1995) footsteps and continue to use the term "science" in this book to refer to observations made by traditional experts about the natural world, but I expand it to include the technologies and methodologies that support it. As

Harding (1998b, 11) argues, "Scientific knowledge is inseparable from the technologies of its production; these have social and political preconditions and effects, and they provide blueprints for subsequent technological innovations." This is important to note since democratizing science movements aim to contest and control not only particular findings but also the methods, variables, and techniques of scientific research that empower experts and marginalize activist perspectives. Although it is difficult to demarcate science from nonscience (Gieryn 1983), it is useful to point out the particular characteristics used to justify scientific endeavors since those are often what DSMs challenge. Science gains its authority and differentiation from other expertise through its particular set of characteristics. It contains a variety of components, including theory, experimentation or observation, measurement, interpretation, and results.

Scientific work is generally regarded as objective and above the influence of the societal actors who fund or create it. However, much research has shown that this is not the case, and that all the elements previously mentioned are shaped by the social context from which they emerge (Longino 1990). This debated perception of science leaves it vulnerable to epistemological controversy and, in many individuals' eyes, also open to influence. Despite the possibility that science is socially determined, it is the basis of technologies that are a fundamental part of policies that determine what is legal and of research that shapes what is considered normal (Russell and Gruber 1987).

Similarity across DSMs occurs partly because the formal rules and norms of scientific institutions are similar across local and national contexts (Merton 1996). Institutional isomorphism describes why institutions tend to grow more, rather than less, similar over time (Meyer et al. 1997; J. W. Meyer 2000). This concept applies to science and therefore generates the rationale why DSMs cross diverse types of knowledge and geographic locale. For example, because of the strength of institutional isomorphism, environmental impact assessments have grown to be used worldwide (Glasson and Chadwick 2005). The biomedical model predominates in health care in many countries (Wade and Halligan 2004). Science becomes routinized, and certain types of evidence are now standard for policy makers. The two cases examined in this book exemplify the contestation of environmental impact assessments and biomedicine. Such globalized institutions or medical frameworks established by science are not always congruent with local context (Lakoff 2007). Such disjunctures often instigate movements that contest their implications.

Goals

Democratizing science movements can have a broad range of goals—social justice, normative change, improved democratic practice, altering public perception, and many others. The unique characteristic of the democratizing science movement is its attempt to change the methods, content, institutions, or political impacts of science. More specifically, activists often aim to change the production of knowledge, public opinion and understanding, or the form of scientific institutions. Generally, the goal is to answer the concerns of citizens affected or codified by scientific language and study by removing what movements perceive to be the false objectivity that insulates science from public critiques, as similarly described by feminist and cultural critiques of science (Haraway 1988; Harding 1998a). DSMs also create new institutions through which nonexperts can have an influence on political decision making and shape the scientific process itself. This does not mean that local knowledge or lay perspectives are sufficient to create effective research outcomes or policy making. These movements instead argue that activist perspectives should be a part of these realms that currently include few or no pathways for such participation.

There are several possible targets of challenge: science, government agencies, corporations, and public opinion, among other topics. The scientific target can be narrow or broad—a particular study that misrepresented local perspectives or a commonly used approach of a particular scientific discipline. This process is not linear. Activists simultaneously challenge the authority of experts and adopt expert strategies. Targeting science is a way to demonstrate the lack of democratic accountability of state institutions on the local, national, and transnational levels. By claiming that the scientific basis of policy making is unsound, movements can find avenues to influence research and consequently change policy. Challenging corporations in this process may be the newest mode of contestation for such movements. It has taken a greater role in activism as many government institutions have become privatized or simply spend more time regulating corporate actors.

Like economic actors who use differing bodies of scientific evidence to push forward their agendas, social movements use science as a way to advance their goals. Industry produces its own research, which is then used as the basis of lobbying strategies, despite the fact that the reports they release are often biased in their favor. Social movements therefore critique such research and generate their own in order to counter the power of private interests in political decision making. In a powerful demonstration of this battle between local knowledge and corporate power, Allen (2003) documents the case of Cancer Alley in Louisiana. There, local communities have worked with experts-turned-activists and activists-turned-experts in order to address toxic dumping. She explains how these individuals identified the biases of industry-generated science and the related rhetoric that has disempowered communities.

Tactics

Democratizing science movements may use what are categorized as traditional or new social movement tactics to create social change on a variety of levels. These tactics include public protest and education, state challenge, identity-based organizing, and others. Their tactics may result in social conflict generated by the marginalization of broad-based populations, such as that found in poor people's movements or grassroots action. There are three main arenas into which they insert their perspectives: scientific production, political spaces, and public dialogue. The first activity is possibly the most distinct to DSMs; however, democratizing science can also play a role in the latter two.

Activists become the cocreators of research projects or engage with experts in critiquing existing science. By adding their perspectives to new scientific projects, they attempt to change scientific methods and norms, instigate research in new topics, and create new conclusions. Democratizing science activists see the importance of scientific methodology. They debate the scientific strategies that have been used to assess the topics with which they are concerned. They point out that certain methods and tools are more likely to support industrial interests, and that other strategies should be used to address their concerns. Lay/expert collaborations are central to developing new scientific techniques and pushing forward the study of new variables. This tactic counters predominant scientific approaches and gives more credibility to the areas of research represented by the movement. For example, Corburn (2005) examined four cases of what he calls "street science" in New York City that attempted to account for local exposures to pollution that were not being adequately addressed by scientists. Like other democratizing science activists, these community members argued that science should be "coproduced." Residents' knowledge of local environmental exposures supplemented governmental risk assessment and advanced regulatory policy. Such collaborations are a distinct and controversial scientific strategy that has impacts on scientific findings, norms, and methods.

Democratizing science activists often bypass political institutions in order to change science and validate lay perspectives. Often this is due to the lack of participatory measures. In fact, they frequently engage in scientific production because there is no other pathway through which they can influence policy making. In the majority of democratic countries today, citizen participation in policy is limited. Therefore, activists often also seek to change political institutions to gain a place in policy making. To do this, DSMs must not only engage in science but also respond to existing political opportunities and attempt to leverage new ones. Movements may do this by forming alliances with political actors and challenging specific institutions or institutional forms. Several scholars have reported on the ways in which movements use science and technical information in advancing the state/society synergy that Peter Evans (2002) argues is important to creating accountable environmental policy. For example, O'Rourke (2002) describes what he terms "community-driven regulation" in Vietnam in which community groups pressure policy makers to regulate pollution better. He notes that one of the points of debate between the polluting company and the community is whose knowledge or representation of contamination is legitimate. In Taipei (Hsiao and Liu 2002), activists leveraged a connection with researchers at National Taiwan University to gain greater credibility for their own experiences with the media and government. These cases demonstrate how governmental representatives are more receptive to laypeople who are represented by or affiliated with experts. The need to engage with government institutions has been part of what instigates DSMs.

In instances where a community or particular group finds little recourse for injustice through existing political avenues, it may also be motivated to protest. Some theorists argue that states "make dissidents" (D. S. Meyer 2002, 13) by limiting democratic participation or targeting a group with particular policies. As a result, groups mobilize to protest a specific policy or to increase their access to state-generated resources. On a broader level, political paradigms can generate protest by marginalizing local concerns. David S. Meyer (2002) argues that political climate shapes activists' perceptions of possibilities for effective protest. There must be an opening in the home government or in a powerful foreign state institution that indicates that protest has the opportunity to be successful (McAdam, Tarrow, and Tilly 2001).

Finally, DSMs may attempt to shape public opinion, often by taking their critique of science to a broader level. This means translating findings into lay language and making them accessible for constituents. This process is also a part of "framing," where social movements reorient the experiences and events of their constituents into a new interpretive schema (Snow and Benford 1992). This schema often connects activist experiences with a social justice perspective meant to initiate and sustain movements. Frames often use preexisting cultural concepts that highlight new connections or concepts (Tesh 2000) and reconceptualize public discourse and political language (Steinberg 1998).

One of the most common frames used by DSMs is the precautionary principle, or the idea that precaution is called for when there is a lack of data about safety (Raffensberger and Tickner 1999). This principle is often invoked when some data have shown that a product or action harms human health or the environment. The precautionary principle was initiated in Europe and adopted by American movements. DSMs use it because of its focus on alternative action in the face of scientific uncertainty. Although the European Union has begun to use it in policy making, it is often seen as too radical a step to implement because it could have broad impacts on production practices.

Why Democratizing Science Movements Arise

Democratizing science movements have arisen largely in response to what Jürgen Habermas called "scientization" (1970), which empowers those who possess expert knowledge while often marginalizing laypeople. Since Habermas conceptualized scientization, many other theorists have been concerned with processes akin to it. Backstrand (2002) looks at the role of science in politics in a different way by examing how scientific knowledge becomes a part of authoritative narratives that infiltrate policy-making processes. A fundamental part of scientization is the technical codification of traditionally nonscientific elements such as culture, bodies, and livelihoods. Drori and colleagues state, "Science and its social authority as a general cultural model [is] spreading and affecting society in diffuse ways, rather than solely or primarily as a means for achieving instrumental or technical goals" (2003, 2).

TABLE 1.2 ELEMENTS OF SCIENTIZATION

Factor	Description
Biased knowledge production	Expert-produced science that does not consider the perspectives of affected populations
Insulation of experts in institutions	Government or scientific institutions that protect research endeavors from public scrutiny
Corporate funding of science	Funding of research by corporate interests that leads to biased findings

Although these developments are difficult to identify, scientization entails three more easily articulated processes (Table 1.2): (1) the construction of biased expert knowledge (Harding 1998a; Haraway 1988), (2) the consequent insulation of this knowledge in institutional structures (Bimber and Guston 1995; Jasanoff 1995), and simultaneous (3) corporate shaping of scientific production. This latter element has grown in importance of late because of controversies that have arisen regarding corporate censorship of scientific production, like the results of pharmaceutical studies (Krimsky 2003). These private economic players also play an influential role in major governmental decision making meant to be based on pure science, such as whether or not to sign on to the Kyoto Protocol (McCright and Dunlap 2003). Industry also produces its own research, which is then used as the basis of lobbying strategies. This industry science and influence over governmental decision making are often used to support the market consumption of a new product, consequently increasing profits.

Some point to the power of scientization in claims that "today's society is based on technology" (Castells 2004), that expert "knowledge is more important even than capital" (Knorr-Cetina 1999), and that, if nothing else, science is growing in influence and size (Drori et al. 2003). Experts and the knowledge they generate play a fundamental role in all aspects of life through a variety of traditionally unscientific spheres. The state has been an institutional stronghold in this process, and scientization of policy is one of the most critical implications of the growing power of science. As science is made pivotal in political structures and processes, expert opinion rises in importance above that of the citizen. At the same time, marginalized laypeople have found the vulnerability of this institutional form. The institutional inability to deal with the growth of science and technology (Ezrahi 1990) means that science-based contestation offers a

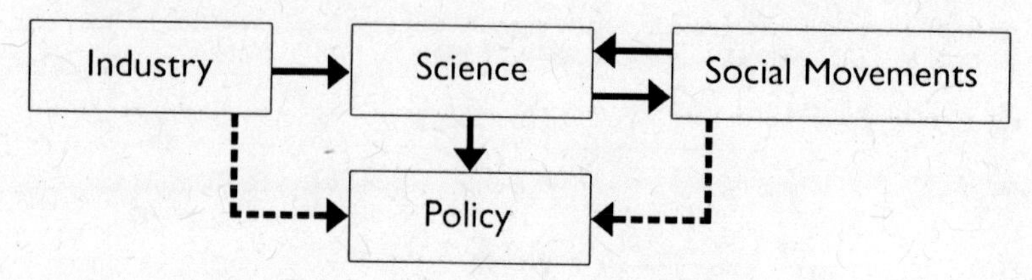

Figure 1.2 Social Landscape of Democratizing Science Movements

gateway for a new kind of democracy in which legitimizing lay knowledge is a path to changing policy. Therefore, democratizing science movements aim to retrieve the power of scientization from experts and reclaim the authority of lay citizens.

Figure 1.2 offers a visual depiction of the social landscape to which DSMs are often responding. Scientization is represented by the two central boxes and the box to the left. This encompasses the processes of scientific influence on policy and corporate influence on science. The box on the right represents the role of DSMs. These movements influence research and are shaped by research. It is important to note that both industry and movements affect policy through other means than just through science. This is represented by the dotted lines. The solid lines represent the main pathways under study in this book.

Scientization is parallel to other social processes like medicalization, the transformation of nonmedical aspects of life into those codified and controlled by medicine (Conrad 1992). In the same way in which people have resisted the transformation of "conditions" into diseases, democratizing science activists argue against scientific misrepresentation of personal experiences and knowledge. In a similar sense, democratizing science movements can be regarded as analogous to advocates who work against technocracy. Such movements have emerged in contexts around the world against "high modernism," or, as Scott (1998) would describe it, development that invests in science and technology, the importance of a "rational" form of order, control of nature, and expansion of production regulated by the principles of science and technology. While democratizing science movements may concern themselves with issues of medicine, bureaucracy, democracy, and other topics, they are distinct in their recognition of the importance of science.

Institutionalizing Science, Marginalizing Citizens

When representation is not achieved in democracy and citizen needs go unmet, movements arise. This is also true in the ways in which movements respond to lack of representation on important scientific issues. Scientific exceptionalism (Bimber and Guston 1995) in policy is one of the most fundamental elements of scientization. It gives science an influential role as the determinant of important policies, as the guarantor of truth, as a sound financial investment, and as more authoritative than other voices. In this sense, as Jasanoff (1996) claims, when political decisions can be made solely on the basis of scientific evidence, concerns about equity and power normally attached to decision making are curtailed. This type of state administration of expert knowledge has led to undemocratic governmental practices and the marginalization of citizens, making the state possibly the most fundamental institution in scientization. Weber asserted that "modern rational capitalism has need, not only of the technical means of production, but of a calculable legal system and administration in terms of formal rules" (1952, 25), resulting in domination of citizens by expertise. Although they do not offer a causal explanation, Drori and colleagues (2003) show that the rise of the bureaucratic state has accompanied a cultural and state-endorsed legitimacy of science. In other words, the legitimacy of science works hand in hand with the legitimacy of the state (Beck 1995).

Some would argue that the orientation of the state to science is useful in that scientific categorization enables states to keep track of their populations and supports industrialization and state building. Part of this argument is based on the idea that transfer of scientific technology from developed to developing nations is also fundamental to increases in output and state building in developing nations (World Bank 1999). This argument supports efforts by many first-world universities and lending institutions to create technology-transfer projects. These projects endorse investment in scientific developments and their wide application across contexts. However, Scott (1998) argues that state projects without "metis," or public knowledge of local context, are often great disasters. Environmental and social destruction resulted from a high modernist approach that denied a dialogue between the "imperial" scientific view and other forms of knowledge. This approach accompanies a larger societal organization along scientific lines of empiricism, enlightenment, and rationality. It has been transferred from developed societies to developing ones and has created the current scientific mode of governmental decision making (Alvares 1992). Such patterns of policy making, based on the imposition of first-world scientific methods and theories on third-world societies, perpetuate a Western model of development and an accompanying set of inequalities (Harding 1998b).

State institutions and policy play a role in scientization through their own institutional power and as an interlocking part of microlevel scientific practices. Cohen and Arato (1990) propose that postindustrial society is characterized by "new loci of power, new forms of domination, new modes of investment, and a 'self-reflective' cultural model" (517) where research and development, information processing, biomedical science, and mass media make social life vulnerable to technocratic control. As a result, "stakes of social conflict revolve around the institutionalization of this cultural model: autonomous, self-governed, egalitarian institutions vs. elite-controlled, technocratically managed structures permeated by realms of domination" (517).

When Are DSMs Co-opted or Successful?

DSMs face an uphill battle, struggling against powerful, entrenched interests that often have been able to purposefully or inadvertently co-opt activism. There are ways to develop participation that accommodates the needs of movements, such as in deliberative democratic political and research institutions. However, truly participatory institutions are difficult to create. The same social conditions that generate DSMs may also limit their ability to achieve change. Limited mechanisms for participation in scientific or government institutions may exacerbate conflicts when institutional forms do not change. The actors whom DSMs challenge are generally more powerful and dominant and may develop a response to movements that defuses movement protest without actually meeting demands. As a result, when DSMs challenge the state and corporate and scientific actors or institutions, they face the possibility of being co-opted. While this is true in all forms of activism, participatory mechanisms are particularly poignant because they appear to meet movement demands (Cooke and Kothari 2001).

In this section, I first detail how social movement scholars have theorized co-optation. I then briefly outline how co-optation has taken place in deliberative political institutions and participatory research. Pulling these themes together shows that research institutes and scientific processes offer critical new pathways for democracy. However, certain conditions are necessary for avoiding co-optation.

Social Movement Theories of Co-optation

Social movement theorists have outlined how movement groups become co-opted. Although these theories are not specifically tied to those of participatory or deliberative measures, they can help articulate how participatory measures can be created in a way that meets movement agendas. A number of theorists have worked on the theory of co-optation (Gamson 1990; Lacy 1977; Seiler and Summers 1979; Taub 1983; Pellow 1999; D. S. Meyer 1993). Most generally, they show that when lay citizens have only a limited ability to participate in government institutions, their interests may be watered down while the institution with which they engage may still gain credibility for providing participatory measures. Past study has demonstrated the steps through which this happens and provides a model of failure with which DSMs can be compared.

D. S. Meyer's (1993) analysis of antinuclear activism described how movements can be demobilized through fragmentation of political action. As activism is incorporated into permeable and resilient governmental institutions, movements lose their politicized agenda and, therefore, visibility. Meyer claims that as a result, activism outside established organizations leads the cutting edge of movement activism. These activists conduct the more radical and confrontational tactics that Pellow (1999) argues keep movements from "selling out" their constituents. Both of these pieces of research have found that movements that get involved in government institutions must maintain a radical position while participating in political institutions.

However, even when movements attempt to maintain a radical approach to the state, they are often co-opted as more powerful interests first accept movement interests and then reassert control over them (Ruzek 1980). Those who co-opt movements can be state representatives or other powerful interests. Ruzek explores how, in the case of women's health activism, medical professionals touted movement innovations as desirable and then subtly pressured them into the form of previously used medical practices. In that case, more conservative health practitioners co-opted the movement agenda. Large funders of activists can also serve a similar role. McAdam's (1982) case study of black insurgency demonstrated that movements lost control of their agendas because of the shaping influence of funders. He argued that the need for financial support often drives movements to seek external funds, which may in turn dilute the movement's goals or change the direction of the movement.

Co-optation is a social process that may accompany social movement formation and often movement demise. Selznick (1953) outlined how co-optation occurs by describing the way in which movement members are incorporated into negotiations and therefore are more easily controlled. Murphree, Wright, and Ebaugh (1996) both specified and broadened this definition. The authors conceived three categories of (1) channeling, (2) inclusion/participation, and (3) salience control. The second of these steps is the most important for understanding DSMs. Rather than inclusion and participation being a facet of social movement success, they can represent co-optation. Challengers can be neutralized when they are made to feel as though they are participants in decision making even though they in fact have no formal decision-making power. This can be achieved by governmental adoption of movement proposals that do not threaten governmental agendas, inviting movement leaders to participate in official conferences or activities, and creating formalized avenues of complaint that replace protest. Salience control involves a similar process but is not necessarily based on participatory measures. It is the process through which movement groups are supposedly appeased, or sufficient interest is shown by the state or entity being challenged to make the movement feel that its interests are being addressed. As a result, the movement shifts its agenda to a new target.

Coy and Hedeen (2005) similarly describe a four-step model that includes (1) inception/engagement, (2) appropriation, (3) assimilation and transformation of movement goals, and (4) regulation and response. In the first step, a movement forms and the target of challenge recognizes the need to change in response. The presence of elites within the movement facilitates this process. Second, the language and terms used by the movement are adopted by the target of challenge and then redefined. As a part of this process, movement goals are institutionalized as activists are incorporated into policymaking committees. An important characteristic of these institutions is the proscribed number of movement representatives, which leads to limited decision-making power. Movement programs are realigned to the goals of the institutions into which the movement is incorporated. Finally, policy changes are achieved, but these are not necessarily the most positive outcome for the movement.

These two models of co-optation processes that have been largely conceptualized with state actors in mind are directly relevant to movement engagement in research and with other social actors who control science. Democratizing science movements may engage in political and scientific institutions or processes, such as a public hearing or review panel, and face the very same barriers to participation. In addition, organizations, individuals, and institutions that fund movement activity and purposefully or accidentally co-opt movements in the process are similar to funders of research projects in which DSMs engage. For example, when a DSM gains funding for a new scientific project but is not able to truly shape its process, methods, or questions, its agenda to broaden or shift expert knowledge may be hamstrung.

Participatory Mechanisms

The shape and role of participatory institutions is critical to whether and how co-optation occurs. Much research on participatory measures has focused on how the individual identity of potential participants affects their desire or ability to be involved (Agrawal and Gupta 2005). Of equal importance, however, are the structural constraints that circumscribe these individuals. Participatory democracies are meant to maintain mechanisms through which citizens can take part in influencing political processes. The weak point is frequently the form through which participation is meant to take place. Citizens often do not have formal decision-making power or are so outnumbered by other social actors that their voices go unheard. They are not involved in establishing research questions. They do not have formal roles in decision making, and there is little projected time to establish trust between researchers and laypeople.

Theorists of deliberative democracy and participatory action research have debated how to create truly egalitarian and participatory projects (Baiocchi 2001; Heller 2001; Fung and Wright 2002; Minkler, Wallerstein, and Hall 2002; O'Fallon, Tyson, and Dearry 2000). Most research has focused on state-generated participatory institutions, such as participatory budgeting (Baiocchi 2001) or habitat conservation planning (Thomas 2001). But governments have had different capacities to accomplish such tasks at state, regional, and national levels (Heller 2001), in part because deliberation is often limited by the government rather than encouraged. Participatory mechanisms developed in Latin America have reflected some of these complexities. In research possibly most relevant to the topic at hand, Lemos and Oliveira (2004) have explored the effectiveness of new, participatory measures for water management, such as river basin committees. They have found that executing deliberation in these institutions depends on the access such institutions have to civil society. These developing management systems are partially made accountable by nongovernmental organizations. The few government forums available for lay citizens to engage in research processes or scientific institutions have been structured in a way that fosters similar co-optation. They have allowed a very minimal level of citizen involvement, enough to superficially claim participation, but not enough to allow nonexperts to counterbalance experts (McCormick et al. 2004).

The most common environmental participatory mechanism in the world is the environmental impact assessment (EIA). The EIA is meant to make development plans accessible and transparent to the needs of civil society and local communities. In Brazil, EIAs are administered by IBAMA (Instituto Brasileiro do Meio Ambiente e dos Recursos Naturais Renováveis), the national environmental agency initiated in 1989 (Viola 1997). In conjunction with IBAMA's state affiliates, it has historically been the main institution with a formal mechanism for nonexpert participation. IBAMA is responsible to the Minister of Environment, who formulates the policies IBAMA administers to the public. Through the environmental hearing process, IBAMA decides whether or not a dam can be constructed. Researchers at IBAMA evaluate environmental impact assessments performed by companies proposing a dam. Many of IBAMA's officials are at least partially sympathetic to movement concerns, and the institution's support has been critical to advancing movement concerns about dam building.

However, as Eve, Arguelles, and Fearnside (2002) point out, the formalized processes of environmental impact assessments allow for corporate influence on the outcomes. Corporations control the EIA and public hearing process in several ways. First, environmental impact assessments are funded by companies that intend to build a dam. Although they are required to hire an independent research institute, researchers in these institutes often inhabit both the corporate and research world, which makes their findings suspect. Public hearings of these assessments are not always held, and therefore accountability that could improve veracity goes uncreated. Ultimately, EIAs are often inadequate, reflecting a lack of effective implementation of environmental law and policy. Corporate funders of large projects may also buy out political decision makers who would otherwise be more objective in the public hearing process. Debates over the effectiveness of the public hearing process and analyses of participatory state institutions suggest the importance of civil society in deepening democracy and improving participation.

Participatory research models show that there are similar obstacles to empowering communities in other research settings (Hickey 2002; Cooke and Kothari 2001). This has been particularly true of community engagements with international nongovernmental development organizations. Rather than using participatory action research (PAR) to realize the needs of the local community, these organizations use it instead to implant neoliberal agendas into local contexts through supposed capacity building, lobbying, or education. This is in itself a form of co-optation. These situations signify the newness of DSM inroads into government institutions, the hesitancy with which officials have responded to citizen interest in guiding research agendas, or their insufficient understanding of how to structure fair and equal participatory measures.

New Forms for DSMs

DSMs often encounter minimal and superficial participatory structures in their attempts to give citizens input in policy making or research. Therefore, they must consciously create new structures that are more comprehensive. The forms of lay/expert collaboration most essential to explaining DSMs are the citizen/science alliance (Brown et al. 2001) and popular epidemiology (Brown and Mikkelsen 1990). The citizen/science alliance is a lay/professional collaboration in which citizens and scientists work together on issues identified by laypeople, thereby transforming the insular nature of expert systems into one that accounts for local experience. The citizen/science alliance has been shown to serve a key role in (1) supporting activism, (2) changing attitudes and practices of scientists and activists, and (3) providing a new value structure to some research (McCormick, Brown, and Zavestoski 2003). This formalized interaction is a form of popular epidemiology—people engaging in lay ways of knowing about environmental and technological hazards and then working with professionals to inform environmental health effects. This process has the potential to permanently influence the methods through which research is conducted because of new ideas introduced by those usually excluded from the scientific realm (Brown and Mikkelsen 1990). Activists' involvement in scientific methodology allows them to use their personal experience as a factor in scientific study and to therefore serve as a bridge between the lay population and scientists.

Although not usually applied in this way, the concept of "boundary organizations" (Guston 1999) aptly describes political institutions that are a pathway for corporate influence, and therefore how lay inclusion in them can be a mechanism of accountability. Boundary organizations are institutions that embody both science and politics. Critical components of boundary organizations are hybrids, or social constructs such as conceptual or material artifacts, and techniques or practices that encompass both scientific and political elements (Miller 2001). This concept usually describes institutions in which there is an interplay of scientific and political frameworks and artifacts. However, corporate interests, and now social movements, often drive the creation of these frameworks and artifacts and thus make boundary organizations a mechanism in the debate between them. Since these institutions provide exchange between social groups, their shape may privilege certain interests over others.

In the following chapters, the theoretical assertions in this chapter are empirically substantiated. The cases therein help test whether new participatory institutions can be formed and the challenges that movements face in making them effective.

2 The Environmental Breast Cancer Movement and the Scientific Basis for Contestation

Sharon[1] began attending a support group after she had a cancerous lump removed from her breast. Every week she traveled to a neighboring county where a group of women met. As her time with the group passed, it increased in size. Other support groups also formed in bordering towns. As the community taking shape among these women grew, they began to realize how many cases of breast cancer were in their area. At that time in the early 1990s, it had been more than twenty years since Happy Rockefeller and Betty Ford had gone public with their mastectomy experiences. Audre Lourde (1980) had published *The Cancer Journals,* in which she drew from her personal experiences to critique the stigmatization of people with cancer. More than thirty years had passed since the inception of radical environmentalism, and a similar amount of time had gone by since contemporary feminism had brought women to protest in the streets. All these occurrences were reflected in the consciousness of Sharon's group as they drew water from the tap and held the glass jar up to light from the kitchen window. As they looked, they wondered if pesticides sprayed in their neighborhood were what clouded their water, and if so, whether they were the cause of breast cancer.

Enraged by the lack of attention to the concentration of cases in their neighborhood, these women went door-to-door asking neighbors

---

[1] The name has been changed in order to maintain anonymity.

if they had breast cancer. Soon door-to-door became neighborhood-to-neighborhood, and Sharon had to form an organization with many volunteers to handle all the information being generated. Knowing little about how to technically codify these data, Sharon's group hired several researchers from a local university to map cases. Only then did Sharon discover that she was not the only one in her area assessing breast cancer. Groups in adjoining counties were overlaying maps of breast cancer cases with histories of toxic spraying and other environmental exposures (Swirsky 2005).

Most of these women were white, well off, and well connected. This meant that they could demand attention and get it. Television shows, research studies, marches, and meetings with politicians were some of the signs of their success. These women have drawn attention to the possible link between breast cancer and the environment in politics, science, and the public eye. Although major economic interests and long-held scientific beliefs thwart our understanding of toxics in the environment linked to breast cancer, more work is beginning to address that question.

These groups, which were the foundation of the environmental breast cancer movement, grew in response to rapidly increasing rates of breast cancer, the predominance of the biomedical paradigm that has found few answers about causes of that rise, and inadequate scientific assessments and regulatory responses regarding potential environmental causes of the illness. Local women began to push forward new research into causes. They also educated local communities, political representatives, and other breast cancer movement organizations about environmental health issues. This course mirrors that of the anti-dam movement that arose ten years earlier in Brazil. The movements were similar in their need to focus on problems in certain geographic areas. While the EBCM faced greater scientific challenges because causation of breast cancer is harder to "prove" (Brody et al. 2004), the anti-dam movement could prove impacts but had difficulty having those impacts considered by planners. The EBCM also targeted multiple carcinogenic sources like water and air pollution, exposures in cosmetics, and industrial point-source pollution, whereas the anti-dam movement focused solely on outcomes of dams. In both cases, political dependence on science and the role it played in shaping public opinion made both movements use it to their ends.

Among environmental breast cancer activists, stories like Sharon's are not rare. Breast cancer transforms a woman's life, she becomes an activist, and over the years, through political connections and organizing, she takes part in collaborations with experts. The movement reflects these individual trajectories on a broader level. It has blurred the boundaries between activism, science, and the state, creating new, sustained attempts to shift science. In this chapter and the following, I describe the pivotal role of science in movements by outlining how initial relationships between experts and activists led to sustained movement activism.

Information sharing between experts and activists can result in new opportunities for "resource mobilization," which is key to movement formation and success. Resource mobilization theory posits that access to external resources shapes movements, including outcomes (McCarthy and Zald 1977). There are debates about which types of resources are most important. Some argue that leadership is one of the most critical elements (McCarthy and Zald 1973), while others put more emphasis on the role of the masses. Networks between groups are also important for movement development and functioning, and particularly faciliate solidarity and leadership development (Oberschall 1973). Expert knowledge is a material and discursive resource leveraged by movement actors. In the cases of the EBCM and the ADM, discursive resources, especially information and disinformation, were important for movement mobilization. As researchers assisted the movements by speaking out for them or educating them about research, they offered an intangible service that was sometimes translated into very real research dollars or policy outcomes. Gaining access to technical knowledge was a way not only to form a social movement but also to help develop more participatory development practices.

Various types of collaborations between laypeople and activists caused three main steps of movement formation to take place: politicization of local consciousness, mobilization, and movement development. These led to a second set of outcomes that will be discussed in later chapters: changing science, enabling activists to reframe issues, and sometimes altering policy. Often the first step in the development of a democratizing science movement is making the community aware of new scientific information around which it coalesces. This initial process was similar in Long Island and the south of Brazil. Science served as a traditional resource for movements by helping the movements themselves form. Thus, the process of scientization that forces movements to engage in science has broad impacts for the way that movements develop and function. These cases exemplify the ripple effect of scientization through movements and, consequently, society at large.

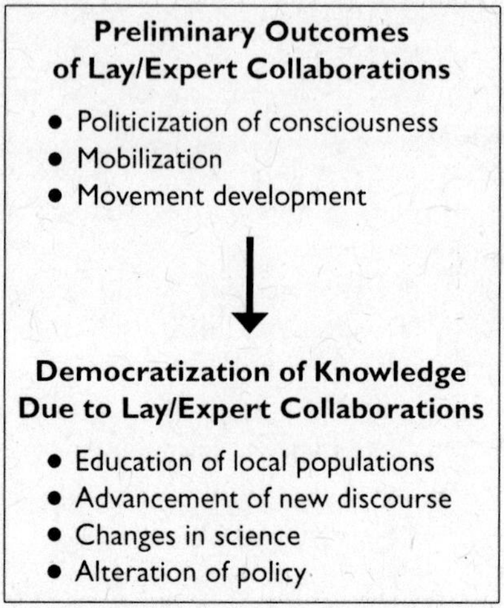

Preliminary Outcomes of Lay/Expert Collaborations

- Politicization of consciousness
- Mobilization
- Movement development

Democratization of Knowledge Due to Lay/Expert Collaborations

- Education of local populations
- Advancement of new discourse
- Changes in science
- Alteration of policy

Figure 2.1 Steps of Lay/Expert Collaborations

This chapter and Chapter 3 describe how lay/expert collaborations directly shape movement initiation. Figure 2.1 demonstrates this process graphically. For example, in one form of collaboration, researchers educate activists. This is often a precursor to activists being able to translate very technical scientific information to their constituents. This type of interaction is a "researcher educator" mode. The "researcher activist" mode takes place when experts are willing to represent community concerns by offering their perspectives and credibility.

Lay/expert collaborations support movement development by politicizing local consciousness and generating mobilization. Politicizing consciousness is often the first step in social movement formation. It involves creating a more political view about an everyday occurrence that can be developed into a movement framework of grievance formation. Social movement theory shows that the construction of a movement identity and movement consolidation are based largely on the politicization of an issue, turning what was once considered a personal trouble into a public issue (Gamson 1992). In this case, the most important aspect of politicizing consciousness is connecting local knowledge with a framework for protest. Lay/expert collaborations provide a key tactic through which local experience is connected with activism. Social movement theorists have often assumed that activism was based on personal identity or a desire to improve rights. While this may be the case, these scholars have largely ignored the mechanism through which an activist's personal experience gets politicized to the degree that she may even be willing to risk her life. DSMs demonstrate that politicizing local knowledge offers a moment in which people connect local experience with a larger framework for struggle. Both the rise of the EBCM on Long Island and that of the ADM in the south of Brazil attest the importance of lay/expert collaborations in movement development.

A Movement Based on Expertise, Lay and Scientific

In 1992, the New York State Department of Public Health released the results of a study that showed that women on Long Island did not have above-average environmental risks for breast cancer. Several local women heard the news and were outraged. One woman, the founder of a breast cancer organization and help line, One in Nine, had been in touch with an increasing number of women with the disease but who did not have traditional risk factors. She explained:

> I said there were only three things wrong with that study—the way it was conceived, the way it was executed, and the way it was reported. I looked at the women who were coming to the support group and they didn't fit the high-risk category. So I thought there's something missing and what could it be? The only thing I could think of was environmental. Everything else was diet, mainly Jewish, old when having had children, or family history. Most people didn't fit that. So the thing we had to look at was where people live and their environment.

She and a group of women contacted two important breast cancer researchers. Dr. Devra Davis was working on developing an innovative scientific hypothesis that explained the link between breast cancer and the environment. In her 1995 article in the renowned scientific journal *Nature,* (Davis and Bradlow 1995) she asserted that unregulated endocrine-disrupting chemicals in the environment would continue to cause an increase in cancers of the reproductive system. Davis and Ana Soto, another well-known breast cancer scientist who had originally found these chemicals leaching out of plastic petri dishes, came to the assistance of women on Long Island.

Activists hoped that with the help of these scientists they would be able to gain funding for scientific study of environmental links to breast cancer. They sought support for their hypothesis that more women on Long Island were getting breast cancer than in other places, and that these above-average rates were environmentally linked.

In 1993, Long Island women worked with these two scientists to marshal backing from other well-known researchers and hold a conference at which the Centers for Disease Control (CDC), the Environmental Protection Agency (EPA), and the National Cancer Institute (NCI) were major presences. After the meeting, several activists paid a visit to the office of Senator Alfonse D'Amato. Their previous event helped get them in the door. With the help of D'Amato and several local representatives, they quickly gained funding from Congress via the National Institutes of Health (NIH) and the Department of Defense (DOD) for several studies of possible environmental causes on Long Island called the Long Island Breast Cancer Study Project (LIBCSP). Breast cancer on Long Island was, and still is, a bipartisan issue, and activists have accessed political connections on all fronts in order to pass new policies and gain political support. Funding for the new study was provided through the first bill in which an activist group stipulated the topic of study. Public Law 103-43 mandated that the following topics be addressed: "1) contaminated drinking water; 2) sources of indoor and ambient air pollution, including emissions from aircraft; 3) electromagnetic fields; 4) pesticides and other chemicals; 5) hazardous municipal waste; 6) such factors as the Director [of NCI] deems appropriate" (House of Representatives 2003b, 1). It provided $32 million in federal funding for study of potential environmental causes of breast cancer. By the end of the project, the total amount spent was closer to $40 million.

Activists in Long Island joined in forming an overarching organization, the Long Island Breast Cancer Network, that united the diverse set of groups that served women with breast cancer who were concerned about environmental risks. This group met on a regular basis both as activists alone and with researchers involved with the Long Island Breast Cancer Study Project during its formation and duration. Local organizations involved in the network included the West Islip Breast Cancer Coalition, Huntington Breast Cancer Action Coalition (HBCAC), the Southampton Long Island Breast Cancer Coalition, Sister Support, and One in Nine, among other groups. Each group had strong political and public clout, and as a network, the groups spanned Long Island.

Researchers addressed a number of variables, both at the request of the community and because of the stipulation of the bill that created funding for the project. They included the impacts of electromagnetic exposures, chemical contaminants in tissue samples, and home pesticide use (Gammon, Neugut et al. 2002). Cancer registries in the area were also established. Many of these studies used new approaches to understanding breast cancer risk. One of the most significant of these projects used geographic information systems (GIS), an innovative computer technology that enables researchers to graphically map multiple exposures. In this case, they overlaid breast cancer cases with hazardous waste sites, industrial sites, or toxic release inventory. They then compared those maps with the layout of drinking-water systems to analyze how toxics might have seeped into drinking water, and whether that seepage might have caused breast cancer.

Collecting New Knowledge, Building a Movement

Lay/expert collaborations were critical to the initiation, development, and shaping of the EBCM. They were a part of the initial attempts to contest science upon which the movement was founded. For instance, when women in western Long Island began noticing that they were surrounded by breast cancer, they began to map breast cancer cases in their community. At first, only a few women were involved in this "lay mapping," working with a map onto which they located cases. After one group was formed that developed mapping for its individual area, it assisted later similar groups in other areas. These activists asked experts from local universities for assistance. Groups interested in environmental causes of breast cancer were then formed with the help of preexisting breast cancer support groups. Ultimately, the concerns raised through their mapping were institutionalized in the form of the LIBCSP. Today Long Island is uniquely characterized by its widely dispersed network of organizations, a configuration of research projects, and multiple small-scale use of geographic information systems (GIS). In this way, the scientific practices adopted by the activist groups directly shaped the organizational structure of the movement.

Even the very mechanisms of scientific projects brought movement concerns into the public gaze and advanced the movement. Lay mapping projects were initiated by Lorraine Pace in Babylon on the north side of the peninsula and Karen Miller in Huntington on the south side (House Committee on Science and Technology 2002). By conducting these surveys, they made people in the area aware that breast cancer was a concern. Both of these women used the attention and credibility they got from these first activities to form organizations that have been pivotal to advancing the overall movement. Their organizations played a role in bringing community members to GIS public hearings, collaborating with other movement organizations either locally or nationally, and, after the LIBCSP was completed, helping initiate new campaigns for improved scientific studies or public awareness. For example, Lorraine Pace made lay mapping an important strategy of other EBCM organizations around the country. Soon after local papers and several national television stations publicized her Long Island project, Marin Breast Cancer Watch asked Pace to come to its area and assist it in similar work.

Working with scientists was also a part of EBCM formation in California and Massachusetts. When the public was alerted to above-average rates of breast cancer on the Cape and in the San Francisco Bay Area, organizations formed and demanded research into causation. Throughout the movement's history, having access to scientific information continued to be a basis for movement organizing. Many organizations hosted events on a regular basis where activists and scientists presented complementary information to conferences or meetings (Massachusetts Breast Cancer Coalition [MBCC] 2006). Often, these presentations were the basis for campaigns that movement organizations were creating. For example, activists in the Bay Area held a series of public hearings on environmental causes of breast cancer that were in part sponsored by the National Institute of Environmental Health Sciences (NIEHS) (Meredith 2002). These public presentations gave them credibility and helped them build stronger alliances with government agencies. They were also a part of what led to larger research projects that supported the EBCM's agenda. When these presentations and conferences began to incorporate government officials or be viewed by the state as valid, they were often critical moments in the movement's history. These instances gave the movement credibility to officials and encouraged movement activists that some progress could be made.

From Cure to Causes: Revealing the Political Implications of Science

After learning about the basics of existing research, environmental breast cancer activists began to critique science as too narrowly focused and based on the biomedical paradigm. Biomedical approaches to disease research focus on individual risk factors, while environmental breast cancer researchers and activists pursue population-level factors (Fishman 2000). The movement argues that the individual-level focus of a biomedical approach omits broader political and environmental factors that might force chemical producers to reduce emissions or to face losses by stopping the production of certain products. An environmental framework looks at changes in industrial production practices as the source of changes in health outcomes. Therefore, the effect of macrolevel structures on individual bodies is more important than individual outcomes alone. The individual is always seen in relationship with the political economy and the social world (Krieger and Zierler 1995). According to Geoffrey Rose's (1985) distinction between studying sick individuals and sick populations, an individual-risk-factor approach seeks to answer the question, "Why do some individuals have breast cancer?" Conversely, a population-based line of inquiry asks, "Why do some groups of women have breast cancer while in other populations it is rare?"

In order to make arguments for a broader political and environmental approach to health, the EBCM must draw attention to less well-known bodies of research that support an environmental hypothesis and address the challenges to such research. Environmental health research is difficult to conduct and often faces scientific critique (Brown et al. 2006). Some of the trickiest issues are lack of data about timing of exposures to toxics, which can play an important role in determining causation; collecting evidence that exposure actually took place, since many chemicals quickly leave the body; and finding a control group, since the American population has ubiquitous exposures to chemical contaminants (CDC 2003). It is difficult to assess the amount of exposure since many chemicals pass through the body and do not bioaccumulate (Schettler et al. 2000). The timing of exposure is important to breast cancer researchers because some research has shown that adolescent girls are more vulnerable than others (Krieger 1989). Some environmental breast cancer researchers have attempted to deal with this issue by creating innovative methods of exposure recall; however, the data found through these methods are far from perfect. These are a number of factors that lab researchers do not face.

Because of various difficulties in conducting research that produces scientifically legitimate answers to lay questions, researchers often use innovative methods and topics of study. These include GIS study, survey methods, and novel epidemiological and toxicological methods. Collaboration with laypeople can also characterize these research projects. Researchers who investigate environmental causes of breast cancer have also used several different types of evidence. These bodies of research include studies of genetics and immigration, pesticides, occupational exposures, and evidence in animal experiments. The diversity of these studies exemplifies the widespread scientific dialogue regulating the legitimacy of claims about environmental contaminants and breast cancer risk.

Critiques of Genetics/Immigration Studies

Much public attention has focused on genetically related research, and many research dollars have been allocated for it. However, proponents of an environmental approach argue that the predominant genetic explanation explains too little and cannot account for the jump in lifetime risk of a woman developing breast cancer from one in twenty (by age eighty) in 1950 to current estimates of one in eight (American Cancer Society 2004). When the genetic defects linked to breast cancer, BRCA-1 and BRCA-2, were discovered, scientists first expected that they would account for some large portion of cases. Since then, the genetics argument has deflated in the scientific world while maintaining public salience. Scientists now agree that genetic causes account for only about 5 to 10 percent of all cases (Davis and Bradlow 1995). Twin studies are generally scientifically respected as a good test of evidence for genetically related disease causation. They reflect a real lack of evidence to support a genetic argument (Lichtenstein et al. 2000).

Similarly, activists have used data from immigration studies to argue that new environmental exposures increase breast cancer rates. Asian women provide one of the best examples. Women who reside in East Asian countries have the lowest breast cancer rates in the world. However, immigration-related data show that the breast cancer risk for Asian women who move to the West increases 80 percent in the first generation, and the rates for their daughters approach those of U.S.-born women (Stellman and Wang 1994). This was known as early as 1973, when some of the first studies of breast cancer rates shed light on this phenomenon (Buell 1973). While some researchers argue that this increase is caused by changes in diet or lifestyle alterations, others claim that increased exposures to environmental toxins are to blame. It is possible that there is a link between the two since pesticides are often absorbed through food consumption. In addition, other types of geographically based studies have found increased breast cancer rates in the vicinity of hazardous waste sites (Griffith et al. 1989; Najem and Greer 1985).

Research on Pesticides and Other Chemicals

Researchers have been examining the link between chemical exposure and breast cancer risk for some time. Generally, they examine specific groupings or breakdowns of chemicals (Dorgan et al. 1999; Hoyer et al. 1999, 2000; Guttes et al. 1998). Wolff and colleagues (1993) showed an early correlation between increased risk of breast cancer and DDE, the chemical breakdown product of DDT, a pesticide commonly used worldwide. This landmark study was the first large-scale environmental research project pertaining to breast cancer that drew major scientific attention. However, later work by Hunter and colleagues (1997), as part of the Nurses' Health Study, cast doubt on their original conclusions.

Since the early 1990s, an increasing number of studies have focused on this group of organochlorine pesticides used particularly since World War II. Dorgan and colleagues (1999) tested blood serum levels of polychlorinated biphenyls (PCBs) for effects on breast cancer risk but were unable to compare different groups of PCBs, as had been done in previous studies, with positive results. They found no support for organochlorine pesticides and PCBs playing a role in breast cancer etiology. However, more recently Hoyer and colleagues (2000) provided new evidence showing the adverse affects of some organochlorines on breast cancer risk and supported a connection to the pesticide dieldrin. Later in the 1990s, Guttes and colleagues (1998) found evidence for a correlation between breast cancer incidence and DDT, DDE, hexachlorocycloethane (HCH), and some PCBs.

In sum, while some studies have shown a correlation between environmental toxins and breast cancer incidence, others have not supported this conclusion. However, many researchers and activists emphasize that lack of evidence does not disprove an argument for environmental causation; rather, it only warrants further research.

Occupational Studies

Occupational studies have more often found support for environmental causation than epidemiological studies (Brody and Rudel 2003). These studies are able to gain a more accurate assessment of chemical releases and exposures. They are often not retrospective, so the data are more easily collected and analyzed. As Labreche and

Goldberg (1997) note, these studies began as early as the mid-1980s and continue to find more positive results than the nonoccupational studies. Hansen (1999) found that for 7,802 relatively young women employed in industries using organic solvents (e.g., textiles, chemicals, paper and printing, metal products, and wood and furniture), increasing duration of employment was associated with greater risk of breast cancer. Among those employed ten years or more in a solvent-intense industry, the adjusted odds ratio for developing breast cancer was 2.0 (Hansen 1999), meaning that on average the risk of cancer doubles after ten years of occupational exposure to solvents. More recent studies show a connection between agricultural and manufacturing employment in which women are exposed to chemicals and increased breast cancer risk (Brody et al. 2004).

Occupational studies may capture information about chemicals that do not bioaccumulate or that dissipate quickly and hence are missed by body-burden studies. However, although occupational studies were conducted earlier and have been more generally supportive of an environmental causation hypothesis, they have tended to receive less attention in the scientific community. Scientists can dismiss them because they represent populations who have faced an extremely intense exposure. The low level of cross-fertilization between occupational and nonoccupational studies is an indication of the fragmentation in current breast cancer research and must be viewed as a major obstacle. In order to create a cohesive body of scientific research that supports an environmental causation hypothesis, it is necessary to combine the results learned from many types of studies.

Laboratory and Animal Studies

Many laboratory and animal studies have demonstrated the effect of chemicals on mammary tumor growth. Many scientists argue that evidence from animal studies is a dependable body of knowledge that supports an environmental causation hypothesis, although there is debate about the ability to translate animal findings to human cases. The first landmark moment in understanding animal mammary tumor development in the lab was a discovery made by Ana Soto and Karl Sonnenscheim (Soto et al. 1991). They were far into a series of experiments examining breast cancer tumor development in cells when they found some breast cancer cells growing exponentially for no apparent reason. Since the research team had not introduced new chemicals to these cells in particular, they were mystified. They ran tests for the known chemical contaminants to no avail.

Soto then remembered that they had begun to order petri dishes from a new manufacturer. When she separated cells in the new petri dishes from cells in old containers, she found the answer to the mysterious cell growth. The new dishes contained a chemical in the plastic, p-nonyl-phenol, that had not been in the old ones. This chemical was leaching into the liquid containing the cells and promoting cancerous growth. This discovery was a historic moment in establishing chemical causation as a credible variable in breast cancer growth, and in advancing the endocrine disruptor hypothesis.

Animal studies have supported the endocrine disruptor hypothesis (EDH), which argues that many industrial chemicals mimic hormones from the endocrine system, consequently disrupting sexual, developmental, and neurological development and possibly causing cancer. It is increasingly considered a scientifically legitimate environmental health theory (Krimsky 2000). A growing number of researchers argue that effects of chemicals are well established, and that it makes sense that endocrine disruptors would affect breast health, which is highly subject to endocrine action. Studies have shown that as animals are exposed to increasing amounts of endocrine-disrupting chemicals, their development changes, and that some animals actually change sex. Colborn, Dumanoski, and Myers (1997) showed that animals and fish in the Great Lakes region have become increasingly female as a result of chemical contamination of the Great Lakes, a phenomenon known as "chemical castration." Some scientists think that these chemicals similarly affect breast development in humans, potentially causing breast cancer.

Scientific Challenges to the Environmental Approach

The scientific community has been skeptical of environmental causation of breast cancer, and proponents have received criticism from their peers and from others outside the scientific community. Krimsky (2000) notes that "scientists have been hesitant about making the association between environmental endocrine disruptors and the risks of breast cancer, despite the fact that the connection between breast cancer and hormones has been established for some time." This is in part because researchers with corporate ties and media sources entrenched in biomedical paradigms often attack this new or less traditional research. Corporate-funded researchers often use the term "junk science" to refer to science that connects human health costs to industrial chemicals. They have written many editorials condemning environmental causation hypotheses (MacMahon 1994; Hunter et al. 1997; Safe 1997), and there have been junk-science critiques of environmental breast cancer research on the part of more conservative scientists.

Hunter and colleagues' (1997) study, considered by many to be the most solid evidence against environmental causation, was accompanied by a deprecatory critique written by Stephen Safe (1997), a breast cancer researcher funded by chemical firms who has been a major attacker of environmental breast cancer research. He began his editorial by using the term "chemophobia" to imply that those interested in chemicals were merely paranoid, and he referred to environmental breast cancer research as "papparazi science." Because this was published in the influential *New England Journal of Medicine*, it carried much weight. The journal's willingness to publish such a diatribe indicates the extent to which established science and medicine will go to discredit the environmental hypothesis.

Conclusions

The environmental breast cancer movement arose when local women heard about research findings regarding heightened rates and began to notice more cases of breast cancer in their area. They saw the need to better substantiate what in the environment might be causing heightened rates. Ordinary citizens in this movement learned science, demanded a seat at the table where research trajectories are determined, and raised new ideas of what constitutes good science. Consequently, research dollars traditionally directed toward understanding individual factors and responsibility in disease causation have been redirected to asking new questions and using new methodologies.

One of the EBCM's tactics is to draw attention to environmental research that has been marginalized in the predominant biomedical breast cancer discourse (Anglin 1997). By pulling these studies together to create a coherent picture of credibility for environmental causation, activists seek to change public understanding and scientific trajectories. The movement continues in a long line of social movements that seek to expand democracy. It does so in a qualitatively new fashion that holds great promise for empowering citizens while at the same time helping improve scientific practices, improve the health of the public, and reshape the priorities of science and medicine.

3 Dam Impacts and Anti-dam Protest

Slight, young, and as blonde as his German ancestors, Marcio[1] is a national leader of the movement of dam-affected people. At home among people of similar ancestry in the southern state of Santa Catarina in Brazil, Marcio does not stand out. Little of his time is spent at home, however. Mostly Marcio is on long bus rides between small cities. His daily activities are always focused on the one goal he has had for ten years, to organize dam-affected people. He moves between populations of affected people, who are generally compact, muscular, and brown from years of farming or fishing in the strong Brazilian sun, and the almost always lighter-skinned governmental representatives who wear suits and whose hands are soft. Marcio dresses the same every day, plain khakis and a white polo shirt bearing the MAB (Movimento dos Atingidos por Barragens, the Movement of Dam-Affected People) logo. This powerful symbol seems to precede and follow him.

At first glance, the MAB symbol is quite similar to that of the Movement of Landless People (MST), which is possibly the most powerful movement in Latin America (Wolford 2003). For MAB, the outline of Brazil is blazoned on a white flag rather than the MST red, inside which is drawn a small figurative picture similar to the MST farmer profiles. MAB's flag depicts a man being crucified on an electrical pole,

[1] Marcio is a pseudonym used to protect the actual individual.

two farming tools crossed at his feet. It is evocative and omnipresent at any march or demonstration. Even the poorest members have been provided with a one-dollar t-shirt to promote the message.

Marcio has been wearing this symbol since the inception of the movement in the south when he and his family protested and halted the construction of Itá Dam. Back then, in the early 1980s, there was no MAB and little critique of dam building in Brazil. Technocrats did not even inform locals when a dam would be built. NGOs proposing the usage of alternative energy generation had yet to be formed. But today anti-dam organizations exist across the country and around the world (Khagram 2004). They contest the accepted model of development by pointing to the inaccuracy, subjectivity, and lack of democracy of the science underlying its planning (Vieira 2000).

This worldwide movement helped instigate an international, independent commission to investigate the negative impacts of dams (Braga 2000) and its consequent study (LaRovere and Mendes 2000). Marcio was a participant in this study through the World Commission on Dams. For that research and other projects, Marcio has traveled to the United States to meet with the World Bank and to other countries to consult with foundations, government officials, and researchers who influence federal-level Brazilian politicians. Marcio has been collaborating with researchers for years. They write books about alternative energy production and, funded by foundations, gather information about otherwise invisible dam-affected people. Through these projects, Marcio and his activist coworkers seek to change policy and make the public aware of dam impacts.

Since Marcio started mobilizing against the Itá Dam project as a young man, the Brazilian government has shifted from military to democratic rule. Trained movement leaders have grown from organizing local struggles to holding government positions. The movement has also evolved from a series of localized uprisings to become a transnational player supported by German, Thai, and South African activists. Marcio has seen the movement stop dams, gain resettlement plans, grieve for the loss of assassinated leaders, and, more than any other occurrence, mourn displacement caused by one dam after another.

Community or local-level campaigns against the construction of large dams, like those led by Marcio, have existed for some time. Driven by the undemocratic nature of dam planning, activists engaged in these battles have found it necessary to engage in disputes over technical knowledge and expertise that are the basis for such policies. They have drawn attention to the biased nature of technical reports and the experts who generate them. In response, they have formed alliances with experts who are open to hearing lay perspectives, and they have developed new research that reflects both technical and lay knowledge.

Protesting Dams, Expanding Science

In the late 1970s, there was an upswing in the construction of large hydroelectric dams, predominantly in the southern region of the country (McDonald 1993). Construction of Itaipu on the border of Brazil and Argentina began in 1975 and ended in 1982, displacing thirty thousand people. The government blatantly disregarded the need for an environmental impact assessment (Roberts and Thanos 2003) and bulldozed its way through local communities, moving people out. At the same time, other dams were initiated. Local communities arose in protest as they were alerted to new construction projects (Rothman and Oliver 1999).

Alliances with researchers were critical in these initial phases of activism, especially because researchers had used information inaccessible to the public. For example, researchers often had information that a dam would be built in the near future and knowledge of what areas would be affected. They imparted that information to local people and ultimately instigated activism. Accurate information about dam impacts was a point of contention. Eletrosul (the southern electrical agency) either did not have information or would not share what it did have with the community. Eletrosul's information was developed to support dam construction and was often inaccurate (McDonald 1993). It included incorrect assessments of potential impacts, the area to be inundated, and who would be affected. Professors served the role of providing accurate information to potentially affected local groups (Rothman 1993). One activist said, "I know that dam-affected people have their own organizations, but they are also subject to a process. The movement was also initiated because of the people from the university and the important part they played in this process."

Access to this information was fundamental to stimulate initial protest. Two local universities, the University of Passo Fundo and an institute at Regional University in Erechim (Universidade Regional Integrada) were key academic groups. One of the first leaders of the movement was a part-time professor at FAPES. Researcher involvement was undermined by university support of other academics who favored development over conservation. When confronted by Eletrosul, university officials did not back their professors but rather attacked them as well. Yet researchers were able to help develop a critique of government claims about future energy demands, initiate community awareness of the dam project, and educate local populations about dam effects (Rothman 1993).

As the anti-dam movement was formalized, activists from the far south who organized the first local organization, the Comissão Regional de Atingidos por Barragens (CRAB), and then the national-level group, MAB, approached researchers at the Instituto por Planejamento Urbano e Regional (IPPUR), an urban studies institute at the Federal University of Rio dé Janeiro. These communities were beginning to organize but had very little political power. There were also no institutions through which they could influence the planning process. When a dam was proposed, they had little choice but to be displaced from their community with no compensation.

Over the past twenty years, these groups have worked together in several ways to alter planning. MAB and IPPUR have affected governmental policy at national and state levels and have laid the groundwork for many other groups to follow suit. IPPUR has been the main partner in collaborative projects with MAB. Experts at the Federal University have advised and educated communities in many regions of the country, have acted as community representatives to government agencies and in public forums, and have worked side by side with affected people to study the ramifications of dam building.

The MAB/IPPUR collaboration has grown in strength and geographic reach since the first days of its inception in the southern region of Brazil in the late 1980s. The past ten years, in particular, have witnessed growth in the number and types of projects. Those projects' collaborative approach has lent them credibility in getting financial support, especially from foreign foundations that want to support grassroots work with infrastructure and guidance. As a result, the group has gained several sources of funding, such as the Federal University, state academic institutes, and the Ford Foundation.

In addition to financial resources, researcher participation with the movement has offered credibility and knowledge. One researcher said:

> I began to teach courses to them about environmental legislation and law and social and environmental impacts of big dams. I had this kind of experience to prepare them for hearings, showing them how it works. We played roles with each person representing different players in the hearing. Also, we made a report on environmental assessment studies. I re-analyzed environmental impact assessments. . . . They asked

> me to support them technically. I call it technical assistance and training courses . . . training materials to educate grassroots organizations on legislation.

The IPPUR/MAB collaboration has taken multiple forms and demonstrates the multiple roles that researchers take working with the movement. Researchers have conducted courses to train leaders to intervene in the political processes underlying energy policy and have advised activists about movement tactics. As activists with and educators of the movement, researchers have helped the movement make political change by contesting inaccurate corporate-generated science and governmental insulation of science. Together, activists and scientists have written new reports and inserted them into governmental and public discourse.

Two such projects have been taking place. First, researchers and laypeople have been developing a resource guide on a variety of energy-related topics, including environmental policy and governmental institutions, for MAB to use in its organizing efforts. Researchers have written some chapters and laypeople others (MAB 2001). The formalization of this knowledge sharing is striking since many MAB representatives are almost entirely uneducated. The resource guide's form legitimizes the perspectives of laypeople and demonstrates the support of IPPUR researchers. Together, the authors of the book developed an alternative way to frame dams and development that counters the state's expert, distant conceptualization of dams. While some chapters detail the specific impacts of dams that governmental sources often deny, other chapters describe local struggles. The authors use this diverse approach to suggest a new paradigm of democracy and development in dam building. The MAB/IPPUR presentation is grounded in local experiences while encompassing an expert understanding of the broader social and environmental situation.

Second, IPPUR and MAB have worked together to generate a database of environmental and social characteristics of seventy to one hundred large Brazilian dams. IPPUR, MAB, the International Rivers Network, and the International Committee on Dams, Rivers, and People (ICDRP) coalesced to apply for funding from the Ford Foundation to support this project. It is an Internet-based resource in which dam-affected people can add their local knowledge to the comprehensive picture of dam impacts. The project is meant to be used most by activist communities that need official information, though academics may also use it. This is a unique instance where the democratization of knowledge means not only the collaborative generation of expert knowledge but also a fairly open and accessible format of presentation in that the entire database will be available online.

However, lack of generalized access to the Internet mean that locals have to go through researchers or other organizations to interact with the database. This signifies some of the limitations to feeding information back to the community, even though communities are effectively contesting expert knowledge in the public eye. Limited engagement with the mass base of the movement has occurred to some degree in the IPPUR/MAB projects. IPPUR researchers have more frequently taken the role of working with the national-level MAB leadership. These relationships have been cohesive and fundamental to movement organizing. The discourse and critiques developed at that level have trickled down to the rest of the movement through organizing tactics.

Since MAB's reach is national, the MAB/IPPUR collaboration is connected to groups all over the country (MAB 2003). After working on these issues for an extended period of time, researchers from IPPUR also are in touch with researchers in many locations. This initial movement has now blossomed into activism throughout the country. In the following chapters, I explore activism in several locations across the nation.

Deconstructing Technical Information, Constructing a Movement

In Brazil, researchers stimulated activism by providing information to local communities about dams that were being considered. Concerned scientists distributed previously unpublicized governmental information and framed dam construction as problematic and unjust rather than an opportunity for economic development, as it is often described. This legitimized local concerns and politicized local perspectives. This politicization countered the broad public acceptance of governmental dam planning. When a community is notified of future displacement, it has the opportunity to coalesce and view itself as a collective identity. In this way, the anti-dam movement transforms the personal experience of displacement into a political issue.

Collaborating with experts concretizes politicization of information through the consolidation of local perspectives into a formalized body of information. As viewed within the framework of my typology, this is particularly true of participation in citizen/science alliances (CSAs). Even in cases where local populations know about a proposed dam and contest it, they do not necessarily connect their own embodied experience or local knowledge to their protest. The personal trouble is powerfully transformed into a public problem when that connection is made. Collaborating with experts provides a space where lay perspectives are legitimized and therefore used to counter official expert discourse. This furthers the initial political framing of dam building by deepening the critique of expert knowledge that underlies policy. When community members assembled in a CSA to analyze and contest an environmental impact assessment, the collectivization of their knowledge helped construct local solidarity rather than the disintegration of community that usually results from displacement. Tucuruí Dam, constructed early in Brazil's history, demonstrates this.

Access to information and the development of a related critique were essential in initiating the anti-dam movement in the south and in later circumstances like Tucuruí Dam. In 2000, the World Commission on Dams (WCD) wrote a report detailing the effects and outcomes of the Tucuruí Hydropower Complex located in the Tocantins-Araguaia river basin in northeastern Brazil (LaRovere and Mendes 2000). Constructed in two phases, the first beginning in 1975 and the second in 1998, Tucuruí has been a controversial project with many environmental and social ramifications. Tucuruí was one of the first major development projects in what was previously a largely untouched Amazonian region. The military government formed in 1964 pushed development in the Amazonian region in order to incorporate the region into the country and consequently make it less vulnerable to foreign interests. Building Tucuruí was a part of this process.

The social impacts of Tucuruí Dam were much greater than originally predicted. An estimated 1,750 families were thought to require relocation at the inception of the Tucuruí project, but actually some 25,000 to 35,000 were eventually forced to relocate. As these people were displaced, thousands of migrants inundated the area in search of jobs, stressing the weak infrastructure there. At the time of construction, Eletronorte gave money to the municipal government to be distributed to the upstream, flooded communities. Those who had been flooded received compensation, while those who lost their access to water did not. These differential compensatory policies divided upstream activists who were partially compensated from those downstream who received no remuneration. Activists reported that downstream communities became more organized than those upstream, possibly because they faced a greater struggle in gaining compensation and so needed stronger organization. The WCD study rekindled interest in the effects of Tucuruí and brought upstream and downstream communities together. As researchers met with and interviewed local communities about dam impacts, these groups realized that there was an international interest in Tucuruí. This gave them the impetus to reignite a struggle for which they had previously lacked support.

Collaboration between scientists and activists was critical in this process. Together, professors at the University of Pará and activists organized a conference in which local communities voiced their concerns and a new dialogue was begun with government officials about indemnification for the communities. Entitled "Popular Movements, Teaching and Research Institutes, and Regional Development in the Area of Tucuruí," this conference provided a space where all stakeholders had an opportunity to share their perspectives. A few phrases on the front page of the pamphlet—"transparency and information for the debate," "popular elaboration on regional development plans," and "science at the service of the people"—exemplified the crossing of governmental, expert, and lay discourses about development at the meeting. The presentations given by researchers from Museu Geolgi and the university, activists from a variety of groups, including fishermen, indigenous groups, *quilombos*, workers, and the movement of dam-affected people, and government representatives from a variety of departments reflect this discursive intermingling. Lay critiques of scientific knowledge that underpins governmental policy were leveraged by several of these groups. For example, activists complained that the governmental codification of an impacted person was incorrect, and that those who had lost access to water were also impacted. By presenting these multiple perspectives together, they constructed a new critique of the Tucuruí project.

The WCD project reflected the first time since the 1980s that groups discussed the inadequate remuneration given to upstream communities. Money had been given to the local government to disperse, but corrupt government officials had kept much of it for themselves. Therefore, this first postindemnification discussion brought these communities together with a common agenda. As local groups got involved in the WCD report, they coalesced more strongly than they had previously.

General governmental and private dam indemnification individualizes the experience of displacement by paying each individual or family for their loss. In this process, broader community or ecological impacts are ignored. The lay/expert collaboration both collectivizes the previously individualized indemnification and recognizes these larger concerns. This happens as communities reflect on their own knowledge of social and environmental impacts. For example, community members claimed that the dollar amounts they were offered for their property were lower than their actual worth. These governmental estimations were generally based on official, codified calculations of land values that did not include the value of ecological and community resources. More specifically, these dollar amounts often did not accurately reflect the productivity of the land. Through recognition of this additional value, official information is contested and local knowledge is counted as a political assertion. As many community members attest to the value of their land, the individual experience of displacement is transformed into a broadly public phenomenon. One MAB document claimed that the organizing provided allowed dam-affected people to think concretely about ways to advance their communities without dams as a part of the equation.

Environmental Science of Dam Building

Like the environmental breast cancer movement, the anti-dam movement seeks to reframe its subject of contestation in order to increase state and corporate accountability and to change public perception. This means highlighting research findings that show the unsustainable nature of dam building. There are two competing paradigms in this struggle: development and sustainability. The first is promoted by governmental agencies, and the latter is leveraged by the anti-dam movement. They have typically been presented in contrast to each other but are not necessarily in conflict. Development can be sustainable, but this has not yet been achieved in Brazil.

The sustainability paradigm is promoted by the ADM. It articulates the need for democratic distribution of economic gain, social welfare, and environmental sustainability. It draws attention to displacement, environmental degradation, and the overall lack of sustainability of dams versus small-scale grassroots development. The sustainability paradigm emphasizes stemming human consumption and depletion of natural resources. It is also based on principles of social justice and environmental protection. In the case of dam building, research that activists use to create a sustainability paradigm draws attention to local displacement, like that of those in the Amazon who are indigenous and therefore officially protected or those in other areas who are also struggling to maintain their way of life. Many indigenous groups have gained reservation territories that are governmentally registered and protected. If any dam project or other development initiative might affect these communities, builders are forced to either gain permission from the indigenous council or pass a special bill in Congress. However, this often does not take place unless activists raise these human rights issues. The anti-dam movement is different from the environmental breast cancer movement in the type of science it contests and attempts to develop. While the EBCM focuses on causes of illness, the ADM attempts to incorporate the outcomes of large-scale development. Unlike the causes of breast cancer, the social and environmental impacts of large hydroelectric dams are well established.

Social Costs

Activists first argue that dams have serious social costs that cannot be remedied by indemnification. Second, they claim that there are critical environmental impacts such as biodiversity loss, the destruction of critical flora and fauna, and greenhouse gas emissions that accelerate global warming. This newer set of concerns has given the anti-dam movement access to powerful global movements and norms about environmental preservation. Hydroelectric dams have displaced over one million people in Brazil. Many populations have not been resettled at all, and those who have been resettled by the state often live in conditions of greater poverty than before they were moved (Araujo 1990). Job capacity often actually decreases as mechanization related to dam building takes the place of more labor-intensive industries. The profits from industry frequently end up in the hands of foreign companies that build the dams. Additionally, instead of contributing to national economic advancement, the billions of dollars lent to the Brazilian government by external sources add to already-large loans, which in 2005 totaled around $300 billion (Fernandes 2005).

The ADM and displaced communities argue that official calculations of dam impacts did not incorporate local peoples' intimate environmental knowledge of and emotional attachment to their homelands. Since 80 percent of Brazilians live in an urban area (United Nations 2001), there is a great deal of rural-to-urban migration in Brazil. However, those who still live in rural areas are strongly rooted in their cultural practices. Being so strongly embedded in the local environment reflects how many families have lived in the same place for generations and hence creates a unique understanding of the evolution of ecological systems. Being rooted in local cultural practices and social patterns also means that displacement has very strong and fragmenting impacts on most groups (Chase 1997). Many dam-affected people expressed great concern about ancestral homes being flooded. A number of studies have demonstrated these impacts in Brazil and around the world. All hydroelectric dams displace populations because of the large reservoirs needed to generate energy. This displacement results in community fragmentation and a sense of displacement that cannot be repaired (Windsor and McVey 2005). Displaced people complained of having no community and being separated from their families. This has also been shown in many other cases of forced displacement when social space, living patterns, kinship systems, and health are negatively affected (Judge 1997). Most studies of dam impacts have no parameters for measuring this social fragmentation or the intimate knowledge embodied by local practices.

Another human impact of dam construction is increased illness in directly affected and adjacent communities. Often, infant mortality rises, and malaria and other illnesses increase at the local level (LaRovere and Mendes 2000). Some of these problems are caused by the fragmentation of community and home and decreased access to proper nutrition, among other social determinants. Others are related directly to environmental degradation, such as standing water that promotes proliferation of mosquitoes that cause malaria. For example, at Serra da Mesa Dam in Goias, schistosomiasis increased dramatically because of dam construction (Thiengo, Santos, and Fernandez 2005). Mercury is also released into the food chain through reservoir leaching and causes illness (Masson and Tremblay 2003).

Arguments about social costs, displacement of rural peoples and destruction of their culture, communities, and livelihood have been part of the basis of organizing around a social justice frame that the anti-dam movement originally relied on almost exclusively. Over time these arguments expanded to include environmental costs.

Ecological Damage

Dam building causes many types of environmental degradation. Dams have affected natural resources, such as critical rain-forest acreage, and natural beauty. For example, the famous Iguaçu Falls on the border of Brazil and Paraguay were once accompanied by a series of celebrated waterfalls that have now been replaced by Itaipu Dam. Compared with other sources of energy generation, hydroelectric dams are in some ways less polluting and more sustainable. Some argue that they produce fewer by-products and pollute less. However, overall, many impacts make them unsustainable. They destroy ecosystems on a large scale, including fish and animal habitats and local plants on which local communities depend for food, fuel, and medicinal purposes (Dixit and Geevan 2002). A number of authors also argue that large hydroelectric dams, especially those in the Amazon, generate massive amount of carbon dioxide (CO_2), which is the most toxic greenhouse gas (Fearnside 1995, 2002b).

A long history of research reflects the ecological costs of dam building. McCully's (2001) extensive review details ecological damages that result from the construction of large dams, such as sedimentary depositing that blocks the passage of water through dams and depletes reservoir areas, decreasing the effectiveness of energy generation. He argues that these unexpected impacts decrease the viability of dams for hydropower, irrigation, flood control, and drinking-water supply. With this perspective, McCully supports the sustainability paradigm and directly counters the rationale of the developmental paradigm.

Possibly the best-researched set of data about the impacts of dams comes from the World Commission on Dams (WCD) report finalized in 2000 (Berkamp et al. 2000). This review analyzed nine large dams around the world in depth for economic, social, and environmental outcomes, in addition to surveying 125 large dams. The report offers negative outcomes on all three fronts and even argues that dams have been more problematic in terms of economic costs than generally thought. Large dams like Tucuruí have been vulnerable to cost overruns and power undergeneration. In almost all cases, loss of biodiversity and habitat and cumulative impacts on water quality were severe. Efforts to mitigate these outcomes have had little success. Wildlife "rescues" promoted by dam-building companies and fish passages, meant to salvage aquatic wildlife, have not been effective.

More broadly, dam planning and construction have long been critiqued for their undemocratic approaches and inefficiency that results in massive costs and less energy generation than expected (Turaga 2000). Historical marginalization of environmental concerns has been accompanied by a disregard for social costs. Some of Brazil's largest dams were constructed before the establishment of its first environmental laws and before the founding of the Ministry of the Environment in the early 1980s. Indeed, that ministry was formed partially a result of discontent about dam building. Unequal access arguments are based on the siphoning of dam-generated energy to development projects nearby or to more urban areas rather than to local populations. This perspective is supported by the fact that industrial sources rank highest for electrical consumption (Santos et al. 2000)

Beyond Movements and Science

The anti-dam movement arose at a time when a military dictatorship offered little information about dam construction and no choice about being displaced. Over the first few years of its existence, it transformed into creating more avenues for influence as the government transitioned to democracy. However, pathways for participation and access to information have depended on the experts who engage with affected people. The early history of the movement and its organizing since reflect this important exchange. Movements can benefit greatly from engagement with science, and science can be greatly affected by movements. This chapter has shown that activist/expert exchange both instigates movement activity and serves as an important source of credibility and information. Without that exchange, neither the EBCM nor the ADM would have been able to move forward in the ways they have. There are also unanticipated benefits of movement engagement with experts, such as the important role scientific information can play in movement development. In an information society, access to information is critical to activists' ability to challenge state institutions or advance their agendas. The ADM made normative assertions that human rights should be respected and that environmental sustainability should be an important consideration in planning. By raising a sustainability paradigm that competes with developmentalism, it has offered a new discourse and set of facts to be considered in planning.

4 Government Institutions and Corporate Interests

Instigating Movement Challenge

One central reason that the anti-dam movement and the environmental breast cancer movement have engaged in science is their need to contest corporate interests that control it. Corporations fund and often shape the environmental impact assessments on which dam policy is based (Fearnside 2006), and consulting companies funded by industry test chemical safety about which EBCM activists are concerned (Cone 2007). Government institutions work in conjunction with these corporations, often by using their research in policy. In addition, governmental research trajectories often fall in line or work in tandem with those of larger, more powerful corporate interests. This chapter describes the corporate/government nexus involved in environmental causes of breast cancer causation and impacts of dams that has instigated these movements. Anti-dam activists have been concerned about corporate funding of environmental impact assessments and the few avenues available to displaced communities to counter the technical reports they often felt were biased or inaccurate. Movement concerns were not just about science, however. Scientific, corporate, and political institutions cocreated public perception and policy regarding dams. As a result, political dependence on expert opinion and corporate control of related technical information drove activists to gain access to science and reformulate it to address their concerns.

In the EBCM, activists have been most concerned about two types of corporate/political/scientific nexuses. The first centers on the complex interrelationship between pharmaceutical companies generating funding for breast cancer research and breast cancer organizations or government agencies that have corporate sponsors with vested interests in individualized disease paradigms. These powerful interests interact fluidly, providing educational material for the public and shaping massive research trajectories themselves. The second and less prominent concern deals primarily with the lack of regulation of cancer-causing chemicals. Chemical manufacturers conduct the analyses of these products upon which regulators depend, and this raises questions whether these policies are protective enough. The EBCM has devoted some time to changing local policies to reduce exposures to these chemicals they argue could be causing heightened rates of breast cancer.

Buying into Biomedicine, Losing the Regulatory Battle

Individualization of disease causation, prevention, and treatment developed in the twentieth century with the rise of powerful polluting industries (Tesh 1988). Miasma theories that acknowledged environmental contaminants and exposures fell away as large-scale polluters were released from responsibility for the health effects of their actions. Breast cancer research has long been focused on the biomedical model of disease causation that uses the individual as a locus of causation, assumes a specific etiology of the disease and the neutrality of medicine, and ignores social, political, or economic facets that drive the development of medical knowledge (Mishler 1981). The largest government institutions, the pharmaceutical sector, and the leading breast cancer foundations have provided support for this approach. In this way, research funding and the tools, paradigms, and methodologies employed by corporate researchers and consequently used by government agencies are "boundary objects" (Gieryn 1983) that create subtle corporate influence on government. More obvious corporate influence is manifest in lobbying by large-scale polluters who stand to sacrifice profits if they are made to better regulate their products (Cone 2007). By influencing political representatives, large private industries have reduced regulation of chemicals linked to breast cancer risk.

The Biomedical Paradigm

Scientific study of breast cancer has increased dramatically over the years. Breast cancer research dollars grew from $90 million in 1990 to $600 million in 1999 (Reiss and Martin 2000; House of Representatives 2003a). The vast majority of this research has been devoted to examining biological mechanisms and individual-level factors such as potential increase in risk due to giving birth for the first time later in life, alcohol consumption, diet, and exercise (Kant et al. 2000; Thompson 1992). Individual-risk-factor approaches continue to dominate for several reasons. First, they fit well with traditional biomedical concepts of disease causation, which are more credible and less politically charged than environmental paradigms. For example, the biomedical model includes a focus on genetics. In a period of growing genetic determinism, genetic makeup has been a major research focus, and knowledge claims based on genetic explanations of disease are considered especially credible (Conrad 1999). A growing amount of resources is being devoted to genetic research at the expense of other approaches. Massive profits stand to be made by those who can offer genetic therapies or own genes themselves. The discovery of the BRCA-1 and BRCA-2 mutations led to much attention to genetic causes. However, it has since been recognized that genetic causes account for only some 5 to 10 percent of all cases (Davis and Bradlow 1995), and that they cannot explain why breast cancer rates have increased in a period that is too short for genetic changes in the population (Davis and Webster 2002).

Second, leading cancer researchers, institutes, and agencies have placed most of their efforts on the biomedical perspective, partially because such research is easier to design, test, and get funded than that of environmental factors. There is also less evidence for environmental causation, possibly because of the lack of scientific tools adequate for environmental health research. Finally, research into environmental causation places responsibility on the business sector and on the government agencies that fail to adequately regulate that sector. The biomedical model supports the interests of industry by keeping prevention and causation focused on individuals rather than broader social, environmental, or political factors. The EBCM seeks to overturn this individual-level approach to health by politicizing connections between health, environment and industrial production, whose interrelationships are a critical part of the social production of disease (Krieger 2003). Breast cancer activists point out that the biomedical model of disease prevention blames women for getting cancer and therefore conveniently removes the responsibility from polluters or government regulators.

Government Funding and Overlaps with Industry

A set of governmental agencies, including the National Institutes of Health and its subsidiaries the National Institute of Environmental Health Sciences and the National Cancer Institute, and the Department of Defense, play the most important role in funding breast cancer research (Casamayou 2001) and in supporting the biomedical paradigm. Additional private funders that take a primarily biomedical approach include pharmaceutical companies, the American Cancer Society, and foundations like Avon and Susan G. Komen, the two largest. These institutions have the most expansive influence on the breast cancer research trajectory through their financial support.

Government funding of breast cancer research comes from several sources. Federal agencies such as the NIH and the DOD receive support from congressional and voluntary giving programs such as the breast cancer stamp, which had generated over $9 million as of 2003 (Breast Cancer Research Program [BCRP] 2003). State agencies such as the California Department of Health provide funding through state budgets and donations on individual tax returns. Nonprofits, foundations, and voluntary health organizations provide the rest. The budgetary allotments of NIH exemplify the lack of general governmental support for environmental research. NIH funds a variety of types of research, predominantly related to genetics, health education, health disparities, scientific training, and disease diagnosis. NIH's one environmentally related institute is the National Institute of Environmental Health Sciences (NIEHS). Out of NIH's 2001 budget of $20 billion, NIEHS received $460,971,000, which was 2 percent of NIH's budget. Meanwhile, in that same year, the National Cancer Institute received $3,249,730,000, or 16 percent, of NIH's budget. Therefore, that year NIEHS' budget was 14 percent of NCI's (www.niehs.nih.gov).

Most cancer research is funded by the National Cancer Institute (NCI). In prevention, the National Cancer Institute prioritizes funding of studies that involve (1) the usage of synthetic drugs such as tamoxifen, (2) lifestyle factors like diet, and (3) preventive breast removal (NCI 1998). The NCI spends most of its resources on treatment options. Its requests for applications (RFAs) often specify the biological mechanisms that it will fund, and refer to environmental toxins only very infrequently. NCI can claim an environmental focus by defining environment broadly and including diet. There are many studies of genetic factors despite the little success that has been gained with such research.

This research agenda may in part driven by the NCI's board of directors. When NCI began to increase its breast cancer research agenda in the early 1990s, the chair of its board of directors was Armand Hammer, chairman of Occidental Petroleum (Samuel Epstein 2005). Occidental later paid the federal government $129 million and New York State $98 million after it dumped 248 assorted chemicals into Love Canal. When locals started to get sick and were moved out of the area, Occidental faced a major lawsuit. This was one of the first critical toxic-chemical cases in the United States (Gibbs 1982). In 1990, $11 million was added to NCI's budget. The following year, the federal government made a major increase of $150 million, with breast and cervical cancers top priorities. The next year that funding increased by another $70 million. In 1993, Congress supported spending another $210 million (Casamayou 2001). In 2007, the NCI's annual budget was $4,791,208,000.

Agencies that fund breast cancer research have often knowingly faced conflicts of interest. Government agencies that support breast cancer research have funded the work of particular researchers who have industry ties. For example, Professor Dimitrios Trichopoulos from the Harvard School of Public Health received a large grant from the Breast Cancer Research Program run by the Department of Defense in 2005 (Clapp, Forter, and Howe 2005). Trichopoulos is a longtime advocate of chemical companies. He was funded by the Chlorine Chemistry Council and testified to the harmlessness of dioxin, which was later banned. Also funded by the sweetener industry, he testified to the safety of its products (Arora 1997). These institutional conflicts of interest in those who drive the breast cancer research agenda often go unrecognized. However, they demonstrate that NCI is a boundary organization through which corporate agendas and interests can intersect with government research.

Because of its corporate relationships, NIH is a similar type of boundary organization. Health effects of chemicals allowed onto the market by the American government are studied by manufacturers themselves or by research consultancies. For example, the Center for the Evaluation of Risks to Human Reproduction is a part of the National Institutes of Health that establishes health risks. Sciences International, a research consultancy, does work for manufacturers of products as well as the NIH (Cone 2007). Although the group claims to have no bias toward industry, recent results of their testing of the plastic chemical compound bisphenol-a have shown otherwise. It argued that bisphenol-a was safe, although recent developments have made it very clear that it is harmful to children's health and may be removed from baby bottles as a result (Austen 2008).

The American Cancer Society (ACS), which also supports a biomedical approach, has faced similar conflicts of interest. Samuel Epstein (1999) claims that ACS has "longstanding conflicts of interest with a wide range of industries, coupled with a systematic discrediting of evidence of avoidable causes of cancer." For some time, its board of trustees included David Bethune, vice president of American Cyanide, which manufactures chemical fertilizers and herbicides. The ACS also has a history of opposing any acknowledgment of possible health risks posed by carcinogenic substances. In 1958, the ACS argued against the Delaney Clause, which regulates certain substances from food that cause cancer in animals. Despite its resistance, the bill was quickly passed and is now an important source of information about risks to human health. The ACS has an impact on government institutions through lobbying. It is one of the largest lobbying charities, in some years spending more than $1 million. *The Chronicle of Philanthropy,* a U.S. charity watchdog, claims that the ACS is "more interested in accumulating wealth than saving lives" (Batt and Gross 1999). Although the ACS is a medical philanthropic institute and not a government agency, campaigns of the ACS and government agencies often work in tandem, with representatives from the NCI and the ACS serving on boards together and as collaborators.

Foundations and Pharmaceuticals

Foundations also provide important support for breast cancer research (see Table 4.1 for details). The two largest of these entities are

TABLE 4.1 FOUNDATION BREAST CANCER EVENTS AND FUNDING

Event (2004)	Money Raised
Avon's Breast Cancer Crusade	$6,514,016
Avon Public Awareness	$5,389,507
American Cancer Society Walks	$32,000,000
Susan G. Komen Foundation Race for the Cure	$96,913,909

the Avon Foundation and the Susan G. Komen Foundation. Both of them claim to be the largest funder of breast cancer research. In 1999, Avon generated $15.6 million through public walks (Leopold 2001). Between 1992 and 2005, Avon raised over $400 million for breast cancer. However, it has faced public criticisms from the EBCM, which has claimed that until 2002, over a third of revenues went to administration, a figure that exceeded the normal amount. Breast Cancer Action (BCA), the Massachusetts Breast Cancer Coalition, and the Safer Cosmetics Campaign have also tried to get Avon to remove carcinogenic chemicals from makeup, with little success (www.safecosmetics.org). These critiques pressured Avon into donating a small sum of money to environmental breast cancer research.

The Susan G. Komen Foundation also raises large sums of money for breast cancer research. Komen funds fall into several funding categories. In 2003, 26 percent went to education, 31 percent to research, 14 percent to screening, 11 percent to fundraising, 11 percent to administrative costs, and 6 percent to treatment (Komen Foundation 2003). In 1996, Komen raised slightly more than $25 million, and in 2003, $154 million, including $88 million from the Races for the Cure.

The third major funder of breast cancer research is pharmaceutical companies. They dwarf both governmental institutions and foundations in the amount of resources they devote to research. There are over fifteen common breast cancer drugs. General calculations about costs to develop a new drug range from $265 million to $455 million (Angell 2004). Breast cancer drugs rank highly in investment by several pharmaceutical companies. For example, AstraZeneca produces four major breast oncology drugs—Arimidex, Faslodex, Nolvadex, and Zoladex. In 2004, Arimidex generated $811 million, Nolvadex $134 million, Faslodex $99 million, and Zoladex $917 million. Total sales of these drugs equal almost $2 billion (AstraZeneca 2004). Annual reports do not relate how much of the proceeds are funneled back into research for new drugs. However, the massive revenues being generated by these drugs exceed amounts from any other funder. Since corporations are driven by a profit motive, their research agenda must be focused on curative drugs that can be purchased by the consumer. Pharmaceutical companies therefore contribute to the scientific focus on treatment and cure rather than prevention.

Lack of Regulatory Measures

Although research agencies and research done by corporations are formally separate from regulatory bodies, in philosophy and approach, these institutions intersect in practice. This reflects what theorists of boundary organizations describe as a fuzzy demarcation between science and policy, as the two realms actually overlap (Guston 1999; Moore 1996). Such regulatory agencies include the Environmental Protection Agency and the Food and Drug Administration, which use the data generated by manufacturers to make decisions about whether a product is safe. At the same time, large corporations are often dependent on government research institutions that produce basic science upon which their research is based. Take the example of pharmaceutical treatments (Angell 2004). Without the basic science most often produced by NIH, pharmaceutical companies would not be able to produce treatment regimens. In this way, boundaries between research and regulatory agencies and corporate interests are fluid and overlapping.

The EPA is charged with regulating exposures to toxic emissions and substances. However, it has often been woefully inept at doing so, if for no other reason than the overwhelming number of chemicals on the market. Over 85,000 chemicals are registered for commercial use in the United States, but only a small portion of them, fewer than 1,000, have been tested for carcinogenicity, and even fewer have been fully and comprehensively tested for noncancer outcomes (Quinn 2002). The EPA does not have the capacity to evaluate all the chemicals that are brought to market, and the regulations in place allow those chemicals to be used although little information is generated by manufacturers of these products. Chemical manufacturers in the United States must test their own products and provide that testing information to the Environmental Protection Agency in order to be able to put a chemical on the market (L. Goldman 1998). The EPA formed the Endocrine Disruptor Screening and Testing Advisory Committee in 1996 to deal with chemicals that potentially especially affect reproductive and developmental health (Hutchinson et al. 2000). Also in 1996, the Food Quality Protection Act was passed to mandate that 9,700 chemicals approved for usage before August 3, 1996, be reevaluated for their safety. The first group of chemicals to be assessed was organophosphates. This group accounts for about half the insecticides used in the United States. Such changing regulations could have protective effects for human health and have major financial outcomes for the industry producing them.

The FDA has a mandate to protect the public's exposure to toxics in foods and other substances, like personal care products. However, its efficacy has been challenged on a number of fronts. Pesticides sprayed onto foods and substances like growth hormones in cattle have been linked to cancer risk (Reynolds et al. 2002). Studies have also shown that ingredients in personal care products are linked to cancer (Colborn et al. 2002). Findings have supported the hypothesis that phthalates in nail polish and parabens in lotions are harmful to health (Hokanson et al. 2006; Harvey and Darbre 2004). However, the FDA admits that it does not have the capacity to keep toxic substances such as these out of products. It states:

> Under the Federal Food, Drug, and Cosmetic (FD&C) Act, cosmetics and their ingredients are not required to undergo approval before they are sold to the public. Generally, FDA regulates these products after they have been released to the marketplace. This means that manufacturers may use any ingredient or raw material, except for color additives and a few prohibited substances, to market a product without a government review or approval. (Lewis 1998, 1)

The FDA depends on manufacturers to test these products. The Cosmetics Ingredient Review Board is the only panel that exists to regulate cosmetics, one subset of personal care products. However, that board is composed of industry members (Cosmetics, Toiletry and Fragrance Association [CTFA] 2003). Some have argued that conflicts of interest on the part of experts who serve the FDA (Krimsky 2003) also lead to lowered standards that favor industry. Salaries of FDA officials are sometimes partially paid by the drug companies whose products they judge. Newspaper headlines explain that the FDA recognizes this problem and is attempting to deal with it. USA Today published one in 2007, "FDA officials criticized for secrecy" (Bridges 2007). Institutional problems and foci have led to the current approach to breast cancer.

Building Dams: Paradigms and Inaccuracies

Like breast cancer activists, anti-dam activists contest several distinct aspects of scientization: biased or limited science, the economic interests that control such research, and the government institutions that make such a research agenda possible. In this case, environmental impact assessments leave out data that are essential to understanding the full range of dam impacts, corporations interested in building a dam fund the environmental impact assessments or misreport results (Fearnside 2002a), and there are few participatory mechanisms through which local communities insert their perspectives into planning. There is a lack of nonexpert representation in the water and energy organizations where government regulations and corporate-generated data intersect.

The present-day situation has historical roots. When the military dictatorship was replaced by a democracy in the mid-1980s (Skidmore 1999), expert opinion began to justify dam planning that had previously not needed validation. Up to that point, several massive and catastrophic dams like Itaipu in the south and Balbina in the north had been constructed, with dire results. Itaipu cost $18.3 billion and displaced thirty thousand people. Balbina flooded an area disproportionately large to the amount of energy it produced and has since been cited as a tremendous disaster in planning (Fearnside 1990).

Slowly, new democratic institutions were formed, including those that plan hydroelectric projects. A series of environmental institutions arose gradually in the early 1980s, culminating in the institution of IBAMA, the national environmental agency, in 1989 (Viola 1997). That same year, environmental impact assessments began to be required for large-scale projects. These EIAs were instigated by the World Bank, which loaned massive amounts of capital to such building campaigns. Pressures on World Bank environmental policy from anti-dam organizers and environmental activists in Brazil contributed to this international change that, in turn, shaped Brazilian policy in the national-level electric agency, Eletrobras, and state agencies (Rothman 1993). Democratic institutional forms were similar to those in Europe and the United States. In those educational settings, officials gain an understanding of the normative role of expert knowledge where scientific studies are the basis of governmental policy, with little room for the perspectives of citizens. The consequent governmental institutions create a lack of accountability.

The Developmental Paradigm

For a long time, dams were symbols of progress and development, signs that a nation was entering a modern age. The famed term for dams—"temples of progress"—coined by Jawaharlal Nehru, the first prime minister of India, exemplifies this attitude. Developing nations followed the trend of the United States and Europe, which began a long phase of dam building in the early twentieth century. Even after industrialized nations slowed dam building, for almost half a century dams went uncontested in the industrializing world. They remained important in the conception of development and were critical to generating massive amounts of energy for growing industrial economies and residential sectors (Goulet 2005). Consequently, dam building is critical to a developmental paradigm (Khagram 2004). This approach rationalizes the construction of dams for energy generation so that transnational corporations can process products, such as aluminum, or to meet national-level urban consumer demands. This often leaves rural communities adjacent to dams without energy and creates a great loss of energy in the distribution process. Brazil receives 90 to 95 percent of its energy from hydropower plants (Ortiz 2002). Increases in international lending during the 1970s oil boom and tightening of oil usage because of price fluctuations (Guest and Jones 2005) led to a need to reduce dependence on oil imports. Importing natural gas or oil, the major energy alternatives, is prohibitively expensive because fluctuation in the U.S. dollar (USD) exchange rate could further exacerbate Brazilian debt.

Government officials and researchers generally regard hydroelectric energy as the most appropriate energy option (Eletrobras 2003). Dams have also historically been considered renewable since they do not literally deplete natural resources and have much lower emission costs than many other forms of energy generation. The Brazilian government argues that dams produce smaller amounts of greenhouse emissions and are more viable than nuclear energy, an option seriously considered by the government in the 1980s (Schaeffer 1990). This conception is justified by the World Economic Forum, which includes dams as a positive measurement in its sustainability index. State representatives also argue that the presence of infrastructural resources necessary for dam building in Brazil, such as turbine and concrete manufacturers, makes hydroelectric dams a better option for national economic development. In addition, studies have shown that hydroelectric dams have the lowest energy cost to build and maintain, can simultaneously provide irrigation and drinking water, and have longer expected plant life than other energy facilities (Smil 2003). There are very few natural resources in Brazil that can be used to create large-scale energy generation, as coal has in the United States.

Most arguments supporting developmentalism, and consequently dam building, are based on engineering science and economic analyses. Although formally charged with analyzing the costs and benefits of such projects, environmental impact assessments are often geared to understanding the most effective way to construct a dam. For example, although alternative methods of energy generation are supposed to be considered in the process of any dam planning, activists often complain that this does not actually take place. One activist said:

> There's this issue of options assessment. You're thinking of building a dam, what are the alternatives? What are the other forms of energy? What are the other ways of building a dam to limit the negative effects? That kind of technical stuff has an effect on planning, but the government tries to ignore it.

Rather, engineers focus their study solely on how to build the most efficient and powerful dam. Therefore, dam planners use a technical approach ultimately geared toward achieving construction. A very small amount of attention is devoted to impacts on biodiversity and communities. This is demonstrated, for example, by one of the first environmental impact assessments generated for the Rio Madeira Dam in the western Amazon. Table 4.2 is an excerpt from that document displaying the segments dedicated to social and environmental

TABLE 4.2 SEGMENT OF THE INITIAL RIO MADEIRA EIA COVERING ENVIRONMENTAL AND SOCIAL COSTS

Ahe Jirau Ocramentio Padrao Eletrobras Estudos De Iinventario Do Rio Madeira		
Outras Acoes Socio-Ambientais	gl	177,177.18
Communicacao Socio-Ambiental	gl	3.847.49
Meio Fisico-Biotico	gl	90,494.44
Limpeza do Reservatorio	ha	42,924.63
Unidades de Conservacao	gl	34,100.00
Areas de Preservacao Permanente	gl	1,500.00
Conservaca da Flora	gl	2,912.109
Conservaca da Fauna	gl	4,646.28
Qualidade da Agua	gl	4,646.28
Recuperacao da Areas Degradadas	gl	409.72
Outros (atividades minerarias)	gl	1,176.78

Source: FURNAS. 2002. *Inventario Hidrelectrico do Rio Madeira.*

costs. This section is one part of a seven-page document that otherwise details the costs of construction. While this document projects that the direct costs of the dam will be $5,965,904, the projected total for purchasing the land, relocating people, and accounting for environmental recuperation is $255,466. This is a very limited amount and, on the basis of how planning often occurs, is possibly an underestimate. Even details of the environmental and social aspects of construction are generated with the goal of sustaining the dam and not the local communities. For example, the largest cost in Table 4.2 is for a conglomeration of environmental works called 'Other Socio-Environmental Actions,' a total of $177,177, whereas the cost of recuperating degraded areas is $409. While the first is essential to maintaining a properly functioning dam, the second is not.

Corporate Interests and Inaccurate Information

Dam building is often executed on the basis of inaccurate or misrepresented research. Anti-dam activists reported that the inaccurate information they received took several forms. When a company initially enters an area to build a dam, it often provides information to the community solely about the potential benefits and none related to problems that might occur. The company usually proclaims that construction of the dam will generate jobs for local people during the building process, and that new tourism like visitors to the dam and boating in the reservoir will create jobs afterward. Although these jobs are often created, they are not localized. Workers come from many other regions as well, and the jobs are temporary. Second, companies offer inaccurate information about the dam itself. For example, they claim that a smaller number of families will be affected than actually are. This is caused at least partially by a limited definition of who is impacted. Officially impacted people are counted only in inundated areas; however, people who are downstream and lose access to water are also impacted. In addition, the reservoir size is often projected to be smaller than it eventually becomes, consequently displacing more people than expected.

The government mandates that a company interested in building a dam must fund an environmental impact assessment. This is a conflict of interest in that the funding source has vested interests in certain conclusions. As a result, these studies are often biased and inaccurate (Fearnside 2002a). Some government interviewees claimed that the accuracy of environmental impact assessments is assured by the mandate that independent research consultants conduct them. However, these researchers move between private consultancies and corporations that propose dams. This framework offers private interests an avenue of subtle influence on policy. State environmental agencies do not have sufficient resources to evaluate these studies. In addition, the co-optation of political representatives and institutions by private interests means that even when a state agency rejects a dam, officials can supersede that decision and approve it, often using the science of the company's environmental impact assessment to justify their actions (Rothman 2001). In this way, governmental agencies both protect the interests of the state, which is focused on developing natural resources in order to create large-scale financial gain, as measured by gross domestic product (GDP), and maintain alliances with private interests from which they obtain personal financial gain.

These institutional practices both lead to and result from historical ties between industry, state interests, and international private funders (P. Evans 1979). These relationships result in the mutually beneficial process of development. Interviews revealed that state representatives often receive major financial gain for facilitating the building of a dam. For example, rather than distributing compensatory resources, officials have kept it for themselves. Even officials within the government attested to this practice. Therefore, even though dams use critical natural resources, political actors co-opted by private interests assent to construction. Despite the immense expense of construction, private investors profit because this investment in construction is far outweighed by the low energy costs once the dam is built. They then use this energy to mine or manufacture goods. In this way, industrial interests dominate the Brazilian electric sector. Researchers, politicians, and activists alike admitted that this was the case. For example, one government official said, "There are economic influences that are very strong in this [dam planning] process . . . those who sell machines and other things, and also those who use the energy after. The large companies have interests. There are a lot of international aluminum interests in this area." With this network of powerful interests, it is difficult for local people to stop the construction of dams, and it is surprising when they are able to do so.

Peter Evans (1979) described this type of transnational involvement as "dependent development," in which a "triple alliance" between local capital, foreign capital, and the state is formed. This development is a process in which core countries became dependent on peripheral elites who lead the growth of an industrialized sector. Massive social exclusion characterizes this process as economic growth results in gain for elites but is not distributed. Meanwhile, international elites became more integrated with one another. This process is still occurring in the electric sector in Brazil, with national and international elites allying with one another. These alliances provide stability for dam builders' massive expenditures and protection from criticisms. One researcher said, "The structural problem is also because of Brazil's major debt and economic policy which is pursuing exports to earn enough foreign exchange to make the payments on foreign debt. Part of this process is that you need electricity for those industries, like aluminum, that need greater amounts of electricity so you need to increase generation and dam building." The $300 billion national debt means that Brazil must make major payments. One of the most viable options is by garnering revenues from transnational corporations that invest in industrial projects in the country. This is preferred to obtaining additional loans from the World Bank, the International Monetary Fund (IMF), or the National Development Bank. The postprivatization structure facilitated connections between transnational and national corporations and reduced dependence on institutions like the World Bank. This structure may have actually exacerbated co-optation of state representatives because their approval is critical to dam construction by private sources.

Many international private interests have played a role in dam building, including the World Bank. One World Bank official said of dams in Brazil, "Even when you're doing God's work you have to perform a cost-benefit analysis." The World Bank largely stopped lending for large dams in Brazil in the early 1990s but reinitiated lending around 2003 (*The Economist* 2003; Environmental Defense 2004). According to officials, the bank has funded irrigation and reserve projects in the desperately dry northeast in the meantime. The role of the bank in advancing sustainability or economic gain is unclear in this case. In other situations, researchers have found that the bank both generates and ignores information about environmental impacts (M. Goldman 2001).

The international energy corporations and banks that most frequently play a major role in funding dam building are from Brazil, the United States, Europe, and Japan. They include BH Billiton from Australia, Tractebel from Belgium, the Inter-American Development Bank, based in the United States, the Brazilian company Vale Rio Doce, the Portuguese Electric Company (Eletricidade de Portugal), and a variety of others. An array of industrial interests, primarily aluminum companies like Alcoa, also fund dam construction in order to use the energy to mine and refine primary resources like bauxite. In 2004, Alcoa invested in dam building, smelting, refining, and airport expansion in Brazil and made $2 billion solely in aluminum (Alcoa 2004). A number of Chinese companies that intend to process bauxite are developing plans with the Brazilian government to fund a new dam in the Amazon (Fearnside 2006; Lemos and Roberts 2008).

Under the privatized electric sector, state and federal agencies regulate dam construction. Transnational lending institutions, international corporations, national corporations, and state agencies fund it. Other national governments like Japan play important supportive funding roles. Table 4.3 demonstrates this principle in the case of Tucuruí Dam.

Even though Tucuruí was built before privatization, foreign sources provided almost a third of the funding for this massive project. It is important to note that private funds from national sources like the Brazilian Bank (Banco do Brasil) are included within the categorization of Brazilian sources. Loans to Brazilian sources mean that the proportion of foreign investment is higher than can be seen here. From the data provided, it is impossible to tell which of these sources provided more of the funds. These data demonstrate the dependence of the electric sector on private capital. This dependence is clear in other projects and companies. For example, in 2003, CHESF (the northeastern regional electric sector) paid 14 percent of its prof-

TABLE 4.3 EXTERNAL FUNDING FOR TUCURUÍ DAM

Source	Total value (thousands of U.S. $)	Percentage of investment
Brazilian sources (Eletrobras, BNH, Banco do Brasil, Caixa Economica Federal, FINAME, resources for equipment acquisition)	2,175,311	72
Foreign sources (Banque de l'Union Européenne, Bank of America, National Bank of Canada, Crocker National Bank, and a consortium of French banks)	833,094	28
Total	3,008,405	

Source: World Commission on Dams. 2000. Report.

its to employees, 24 percent to the government, and 62 percent to private investors, by far the largest proportion (Companhia Hidro Elétrica do São Francisco 2004).

Challenging Corporations

The cases of the two movements discussed in this book are very different in the way in which the state regulates industry; however, they still maintain similarities. The EBCM struggles with the corporate entities that shape breast cancer activism and research, keeping it focused on cure and treatment. The ADM contests corporations that build dams or develop EIAs. Documents from MAB or an EBCM organization reflect how these movements argue for the need to better regulate industry in order to improve health or achieve social justice. Activists connect the need to decrease breast cancer rates or generate electricity sustainably with an attack on the industrial sources they claim are damaging their lives. They argue that government agencies have not been effective in this task, and they therefore target both industry itself and the state that regulates it. The challenge to corporate sources is demonstrated in the introduction of a book written by the activist and research collaboration called Energy Working Group. This international lay/expert collaboration represents all the NGOs dealing with issues of sustainable energy production in Brazil. The introduction states:

> The international economic model has implanted domination for many years, concentrating wealth in the hands of a fraction of the population and condemning the majority to a situation of miserable poverty. . . . [It] treats nature as disposable, following a model of production and consumption in a permanently expanding market, and providing for external interests and reproducing elitist corporate domination (Ortiz 2002: 17).

This is much like the contestation of corporate practices by the EBCM. The EBCM's particular perspective is demonstrated in congressional testimony from Prevention First, a coalition of health-related organizations that promote the precautionary principle, regarding direct-to-consumer advertising:

> Everyone—patients and doctors—ends up relying on the industry's information to make decisions and recommenda-

> tions. And industry's information is not balanced information. . . . It is simply untenable that pharmaceutical companies can mislead the public with impunity, knowing that, by the time the FDA is both called upon and able to act, the damage will be irretrievable (Prevention First 2003).

In this testimony, this coalition of groups that involves multiple EBCM organizations makes a clear connection between corporate power and corporate research. These quotes show that the EBCM and the ADM attack corporate expansion and control with the intention of empowering the public and their constituents.

Both movements also aim to counter corporate sources and the expert knowledge they produce by constructing new participatory mechanisms and valuing the knowledge of those affected by corporate practices. One MAB document detailing a proposed revision of the electric sector claimed that the new sector should be "democratized with the organized participation of affected populations in planning, decisions, execution and policy-making," and that there should be "events that profoundly deepen debates and involve multiple sectors of popular movements who have different experiences, knowledge and perspectives (MAB 2005)." Another document that emerged from the 1997 international meeting of dam-affected people claimed that "a most important issue is that of the role of traditional technologies of water management, which were serving and even now serve very large areas all over the country. Many of these have fallen into disuse and are neglected, but offer vast hope."

Similarly, EBCM organizations maintain that women's own knowledge of exposure and their lay perspectives on science are of great importance. For instance, the Massachusetts Breast Cancer Coalition works closely with researchers at Silent Spring Institute in order to offer the embodied knowledge of women with breast cancer in advising research. As in the valuation of traditional water technologies, the MBCC and Silent Spring claim that the knowledge of women with breast cancer is uniquely valuable in breast cancer research. In order to give that lay knowledge a place to shape research, Silent Spring has a Public Advisory Committee composed of activists from MBCC and other EBCM organizations. An excerpt from another organization's newsletter shows this valuation: "By including advocates in all federally funded research projects, we will no longer be 'leaving researchers to their own devices,' but will be moving forward as a team to address the critical health issues of our time (BCA 1993)."

Science is a nexus of power between corporate and state interests. Science funded by corporations and used by the state embodies the intersection of powerful interests manifested in technical outcomes. Because the state maintains an image of neutrality on such scientific results, these movements have often moved to directly confront the corporate entities that have clear vested interests by introducing lay perspectives. In this way, they attempt to gain credibility that is often challenged by more powerful interests. A letter written by MAB to the Inter-American Development Bank, which funded a large dam it protested, provides an example of this contestation process:

> [The company claims that] the occupation of the Cana Brava dam work site by the affected populations was "totally illegal and without legitimacy." The occupation of the Cana Brava work site, which took place earlier this year, reflected the disdain with which the company has treated the dam-affected populations. Taking over a work site is a last option used by populations to pressure the company so that their rights may be respected by the constructors and by responsible authorities. This civil disobedience was carried out by the dam-affected families themselves—men, women, and children, mostly small farmers, who in despair face the imminent loss of their way of life and the certainty that they are not being fairly compensated for these losses. Therefore, it is a lie that this occupation "was conducted by major landowners . . . aiming at increasing the price of land at the reservoir area for their own benefit." Once more, the company uses false statements to try to discredit the legitimacy of the mobilization to cover up its own disrespect for the rights of the affected populations (MAB 2000).

Environmental breast cancer activists have similarly challenged industry when they see no governmental mechanisms through which they can achieve their goals. This is particularly true in their Campaign for Safe Cosmetics. A coalition of EBCM organizations and others, including BCA, MBCC, and Huntington Breast Cancer Action Coalition, pushes cosmetics companies to remove potentially hazardous ingredients such as phthalates in nail polish and parabens in lotions from their products.

> The chemicals in any one consumer product alone are unlikely to cause harm. But unfortunately, we are repeatedly

> exposed to industrial chemicals from many different sources on a daily basis, including cosmetics and personal care products. Many of these chemicals have gotten into our bodies, our breast milk and our children. Some of these chemicals are linked to cancer, birth defects and other health problems that are on the rise in the human population (Campaign for Safe Cosmetics 2005).

Through the efforts of the coalition, some companies like Estée Lauder and Procter and Gamble have voluntarily removed hazardous ingredients, while other companies have refused.

Working on multiple fronts in a variety of locales, these movements have the capacity to change the practices of such institutions by introducing new lay knowledge and perspectives. While the EBCM links the environment to health and the ADM relates it to social exploitation, both use these connections to demonstrate inequalities caused by economic interests that dominate environmental decision making.

Conclusions

Relationships with economic interests shape governmental paradigmatic and regulatory approaches. Science is a mechanism used to legitimate these approaches. Often, this legitimation develops within a boundary organization where politics and science intersect. In addition, researchers whose funding or affiliations lead them to side with industry are also critical in these settings (Bekelman, Li, and Gross 2003). These institutions are often the site of policy decisions or at least develop recommendations for policy making. In environmental and state agencies in Brazil or NIH in the United States, corporate interests and government representatives exchange boundary objects, such as EIAs and findings about chemical toxicity. In essence, they coproduce policy by sharing scientific norms and tools that are the basis of decision making. Activist organizations attempt to insert their agendas and scientific findings in order to disrupt trends in policy making. They challenge large-scale financial and government interests to do so.

Because of the immense influence that industry has on science, movements must learn and then critique the science in order to change powerful paradigms. Although the types of science and technical reports being contested are quite different, there are several social and scientific processes that make both the ADM and EBCM, and possibly many other democratizing science movements, contest science—limited participation in research and regulatory policy, inaccurate and unbalanced science, and corporate influence on scientific development.

The types of relationships between government representatives and corporate sources and the openness of state institutions to movement demands have made these movements adopt slightly different approaches to democratizing knowledge. EBCM actors felt that they definitively needed to push more research toward environmental causes in order to shape governmental decision making. The ADM knew that democratizing knowledge was a critical tactic, but directly opposing dam building through protest was similarly important. The different levels of emphasis on democratizing knowledge were partially related to the level of state/industry alliances manifested in the usage of expert knowledge. In the case of Brazil, movement actors have been more directly pitted against private interests since the inception of energy-sector privatization in the mid-1990s. At that point, private companies that intended to build a dam began to be in charge of commissioning the environmental impact assessment. Therefore, industry had a serious influence on what kind of research was generated on environmental impacts. This process of privatization and its associated methods of expert knowledge production weakened the movement. The movement consequently had to engage in direct tactics of political action, like sit-ins and protests, that would physically prevent construction. Although EBCM actors have similarly opposed private corporations, they do so less since the scientific points they need to make are harder to prove. Consequently, they challenge corporations only after they have gathered enough information to show that corporations' actions are harmful or when they find an inadequacy in judgments regarding exposures. Both movements use the embodied experience of their constituents to contest corporations and industrial interests and to introduce new discourses about their subjects of contestation. At the same time, lay/expert collaborations shape activism broadly defined, as well as the experience and knowledge of individual activists. The following chapter explores in detail the lay/expert collaborations generated by each movement and their outcomes.

5 Democratizing Science

Kari[1] grew up in beautiful Washington State in a small town down the street from the Hanford Nuclear Site. Every day as she entered school, Kari walked by a poster of a mushroom cloud, a testament to the power and danger of the facility nearby. Some afternoons when she was alone in the house, Kari would answer the phone to hear the Hanford nuclear emergency test phone call shouting, "Red alert! Please stand by for roll call!" This monthly test and the nuclear threat were a part of normal life, much as the growing number of cancer deaths was becoming routine to people across the country. When her father died of cancer at age forty-eight, the hazard that had seemed in some ways so far away became closer to home than Kari had ever imagined it would.

In her first job after college as an environmental reporter and then during her graduate work in environmental psychology, it became increasingly clear to Kari that more often than not the causes of environmental problems and cancer deaths were obvious. But getting people to take care of the obvious problem was a political challenge. Often the powerful interests that caused the problem would attempt to make the threat seem like a far-off hazard. She began to learn more about how people psychologically distanced themselves from the increasing number of risky technologies that they

[1] The name has been changed to protect anonymity.

encountered every day. She saw how individualizing environmental protection by putting on safety gear made this distancing process easier and hence slowed political action that at-risk groups might take. She saw that both this public perception of risk and the way that science portrayed risk were pivotal to either generating or slowing political change. So when Kari became the director of a project where researchers and activists together examined environmental causes of breast cancer, it was no surprise. She had chosen to become a part of making research an avenue to political action.

Around the same time Kari formed her research team, Linda[2] was diagnosed with breast cancer. She was forty-two. Like most women when they are first diagnosed, she was no activist. She was a doctor practicing psychiatry in Boston. But the mental challenges she underwent after diagnosis—denial, grief, anger—were hard to handle. One day, soon after she began to accept her diagnosis, Linda sat outside in the famous Faneuil Hall area of downtown Boston. It was a typically cloudy summer day. What made it different were the hundreds of women dressed in bright pink shirts carrying placards and banners on the street below. This was the first major breast cancer rally in Massachusetts, the moment in which breast cancer became publicly known as an epidemic in that state.

Drawn in chalk on the black concrete was the shape of Massachusetts. The rally organizers asked all women with breast cancer to come out of their seats and step into the map. Remembering, Linda said:

> So many people came out of the stands, I couldn't believe it. Young women, old women, black women, white women. I was flabbergasted. It was at that point that I said to myself, "When I went to medical school breast cancer was rare." I'm 42. I don't have a family history. I suddenly have breast cancer. I'm looking around at hundreds of women who have breast cancer and this isn't rare.

That was 1991, the same year in which she joined the board of the Massachusetts Breast Cancer Coalition. The next year, in 1992, the Department of Defense (DOD) initiated the Breast Cancer Research Program, run under the Congressionally Directed Medical Research Programs (CDMRP). True to its title, this program was created so

---

[2] The name has been changed to protect anonymity.

that Congress could mandate certain types of health research. In 1993, Congress appropriated $210 million to peer-reviewed breast cancer research. This type of research, which requires that proposals be evaluated by a panel of experts for their merit, was not unusual. What was striking was the inclusion of women with breast cancer on review boards. Activists had demanded that they be able to influence what research was done, and Congress had mandated that advocates sit on panels that reviewed DOD-funded scientific proposals. The Breast Cancer Research Program (BCRP) was meant to involve "consumer" advocates with clinicians and scientists to identify gaps in research and develop new scientific projects. These individuals were meant to guide the funding of the program. This was the first time in history Congress had mandated that nonscientists have the power to influence research. In the following ten years, Congress appropriated a total of $1.4 billion for research on breast cancer and other illnesses through this program (CDMRP 2003).

In 1996, Linda, along with MBCC member Anne Perkins, became one of the first to serve on the DOD peer review panels. After being selected that year, the two women were sent ten to twenty proposals each and asked to submit detailed written comments on them. This was no small task. Even though Linda had medical training, she found it challenging. Once she reviewed the proposals herself, analyzing the topic, methods, and overall relevancy of the research, Linda met with the whole advisory board, whose members sat in a room together, debating studies and deciding their fate. The group met for prolonged periods of time, generally a long day or weekend where its members would enter a room first thing in the morning and not leave until the sun had long set and their eyes were watering with fatigue. Together, their evaluations determined what the DOD would fund.

As the theory of boundary organizations explains the intersection of different actors within an institutional setting, the concept of "boundary movements" explains how laypeople and experts cross traditional boundaries between movements and science in such ways (McCormick, Brown, and Zavestoski 2003). DSMs are often boundary movements whose actors cross the traditional definitions of activist and also the clear boundary between science and nonscience. The lay/expert collaboration in particular can represent an arena of mutability for scientists and activists to move between scientific and lay worlds (McCormick et al. 2004). They occupy space between traditional social roles. This fluidity allows professionals to play varying positions over time, occasionally being part of the movement as either members or "advocacy scientists" (Krimsky 2000), at other times being somewhat detached scientists, and at other times being uninvolved. Fluidity is similarly represented in Steven Epstein's (2001) notion of analytical blurrings. He asserts that we can no longer adhere to binary sets, such as insider/outsider and lay/expert, because of the flexibility of such shifts among both individual participants and organizations. It is important to note this duality of researchers and laypeople because it signifies a shift in what has traditionally been seen to be the biggest obstacle in participatory research—the gap between layperson and expert.

Most of this chapter is devoted to looking at the citizen/science alliance (CSA) and the collaborative forums that are created specifically to change science and policy. Engaging in science can politicize the consciousness of communities and consolidate formerly fragmented groups. Such collaborations contribute to movement development, offer an alternative frame to traditional scientific and political approaches, and also change science itself. The changes in science that result may seem the least significant outcome in that they do not necessarily lead to immediate improvements in the lives of people outside the lab. However, new tools, hypotheses, and data can often be the first steps in radically altering real-life experiences. The most fundamental scientific changes that resulted from CSAs were (1) the broadening of the definition of what should be included in studies and (2) the inclusion of new information provided directly by community members as scientific knowledge.

Although many of these outcomes are positive for both movements and science, these types of collaborations often make movements vulnerable to co-optation because they demand direct interaction with the institutions the movements challenge. Some conditions are necessary to ensure the effective functioning of the institutions that formalize participation. They include (1) formal decision-making power for laypeople, or at least the capability to influence decision makers, (2) the ability to add lay knowledge to studies or presentations, and (3) involvement of the affected group from the commencement of the project. These are general rules and guidelines that work across case or context.

Typologizing Collaborations

By combining theories of participation and social movement research and assessing the cases in this research, I have developed a typology of collaboration that spans national context (see Table I.1).

This is an important part of the theoretical toolbox for understanding lay/expert collaborations and democratizing science movements. It is used to compare the collaborations in these two movements. Collaborations involve four main components. First, in the researcher educator form, researchers serve as educators of movements by providing data, technical information about study methods, and training. Second, in the researcher activist form, researchers act as movement leaders or political representatives. In those instances, researchers offer their scientific credibility to laypeople by speaking or acting on the behalf of movements. Third, in the citizen/science alliance, researchers and laypeople work side by side to construct new research. Fourth, in the collaborative forum, laypeople and experts officially comment on existing science and policy about the subject of contestation in a governmentally sponsored setting. The first and second forms alert affected populations about issues of concern and initiate or support activism. Researchers also use these forms to increase the movement's legitimacy and political know-how. In the third and fourth forms, lay knowledge is translated to a more official level.

Much of the greatest scientific change that also leads to shifts in policy takes place in the citizen/science alliance. It has not only altered the attitudes and perceptions of participants themselves but has also created ideological changes in science that ultimately affect policy. Activists use their knowledge of environmental and social impacts to argue for increased accountability of science and governmental institutions. Often, these groups critique existing science by arguing that experts accidentally or deliberately miss information. Consequently, they insert a new normative structure into scientific assessments by claiming that other variables should be accounted for more completely. As lay knowledge is inserted into expert systems, expert language is opened to value the substantive content and normative differences of lay knowledge. In this fashion, power flows inherent in the microlevel practices of research and science are reversed or made more democratic. Population- or community-level state-generated knowledge is then democratized to include formerly marginalized views.

When Does Participation Work?

Scholars of participatory research have argued that there are specific conditions under which these projects work best (Cooke and Kothari 2001). As previously mentioned, however, these scholars are often divided by national context and consequently deal solely with the developed or developing context. Identifying the conditions under which DSMs effectively engage in CSAs shows that similar circumstances enable effective participation across the globe. Examining the goals of community-based participatory research (CBPR) and participatory action research (PAR) helps articulate the differences and similarities of these approaches.

CBPR has often been the paradigm used in the industrialized world to assess participatory research. Some of the main characteristics of CBPR include (1) power sharing on the part of the researchers with local people, including who defines the research problem and then analyzes, owns, and acts on the information (Cornwall 1995); (2) building on strengths of the community, such as skills, networks, and social capital (Israel et al. 1998), and allowing community members to take part in data collection (Northridge et al. 1999); and (3) building mutual trust and cooperation (Schell and Tarbell 1998). As Steven Epstein (1995) argues, these characteristics allow activists to weave between political, ethical, methodological, and ideological arguments. The main purpose of CBPR is generally to conduct a new study that makes a scientific contribution.

However, communities often care more about generating new knowledge and using it for action or community empowerment. This is more often the focus of PAR, where researchers and laypeople create knowledge in order to have an effect on people's daily lives, with a secondary intent of affecting the trajectory of research (McTaggart 1991; Pain and Francis 2003). While the EBCM reflects the CBPR approach and intent more specifically, and the PAR approach better describes work with the ADM, both types of participatory research function best under conditions of formalized participation, long-term engagement, and conciliation of researcher/activist agendas. These conditions facilitate resolution of scientific conflicts, as is reflected by successes and failures of DSMs.

In DSMs, movement organizations or leaders are often the initiators of research projects. They might first be made aware of an issue through the distribution of statistics or technical information, but in many cases they pursue a research agenda afterward. Because they are the initiators, these projects often do not face several of the traditional challenges or critiques of participatory research, such as community members getting involved only after the research is designed or the research agenda being solely driven by academic interests. Community members are generally involved from the beginning, which means that research questions and goals have a higher likelihood of being shaped by those most affected. Possibly even more important, and distinct from other types of participatory research projects, is that most often data are collected and research is developed with a specific social movement agenda in mind. For example, research could be aimed at passing a particular policy or altering public perception of a topic. When research is initiated by activists and its agenda is to address local problems, there is a greater likelihood that trust will be developed between laypeople and experts and that participation will meet the needs of both researchers and activists.

New Scientific Questions and Tools: Broadening Perception of Causes

When breast cancer activists first began instigating participatory projects, a few sympathetic researchers were cautious but engaged, and more generally projects were met with skepticism. Over time, however, movement groups gained more credibility in research projects, and activists helped develop new research tools, methods, and topics of study. By doing so, they generated new understandings of what might be causing breast cancer. Consequently, the EBCM has instigated several effective participatory collaborations. While Long Island activists were the first to take a scientific tack, the activists who followed soon after in Massachusetts were able to achieve greater scientific innovation and deeper participation. Researchers and activists from the Massachusetts Breast Cancer Coalition formed Silent Spring Institute to develop new science that would investigate potential environmental causes of breast cancer on Cape Cod, Massachusetts, in particular. Through innovative epidemiological and participatory methods, they have developed new science that completely reframes debates about breast cancer. This scientific team has also been a resource for EBCM activists around the country.

Silent Spring Institute

Cape Cod, Massachusetts, has a history of breast cancer incidence that is 20 percent above the state average (Silent Spring Institute 1998a). A few other areas in the state, like Newton and Marblehead, also have heightened rates. EBCM activism grew up on Cape Cod and in the rest of the state in order to address these rates. Activists founded the Massachusetts Breast Cancer Coalition (MBCC) in 1991.

MBCC focuses its activism on creating awareness of environmental causes of breast cancer by educating the public and pushing for related research. In addition to MBCC, the other main Massachusetts group is the Women's Community Cancer Project (WCCP). WCCP has focused primarily on a feminist analysis of all cancers. It has engaged in public protest and education, including a large-scale mural in Harvard Square. The more radical nature of this activism includes a critique of companies that produce both carcinogenic chemicals and cancer drugs, what groups term the "cancer industry." They also point to the need for corporate responsibility and critique direct marketing of cancer drugs.

In conjunction with scientists, MBCC formed Silent Spring Institute (SSI) in 1994 through funding by the passage of a bill in the Massachusetts legislature that provided $1 million a year for breast cancer and environmental research. The plan for this research included activists and scientists working together to develop research on possible environmental causes of breast cancer and educating the public about them. The first phase of study was completed in 1997 with four main accomplishments: (1) development of the geographic information system (GIS), which enables researchers to map homes of diagnosed women and compare women with environmental data (summary maps may be viewed at www.SilentSpring.org, while confidential information about individual women is protected); (2) historical study of pesticide use and drinking-water quality on Cape Cod; (3) development and application of new field-sampling, chemical analysis, and bioassay methods to study environmental estrogens; and (4) detailed surveillance of breast cancer incidence (Silent Spring Institute 1998a).

Phase two of the project, which is completed and whose results are just beginning to be released, involved interviewing 2,100 Cape women to identify both individual-level and community-level environmental risk factors. Interviews have been completed, and air, dust, and urine samples have been collected from 120 homes. Researchers examined home use of pesticides and wide-area applications for gypsy moth and mosquito control, agriculture, and other purposes in conjunction with established breast cancer risk factors such as family history of breast cancer, reproductive history, and use of pharmaceutical hormones. They also developed new measures of possible endocrine-disrupting chemicals and mammary carcinogens in households (Rudel, Brody et al. 2001; Rudel, Attfield et al. 2007) and estimates of environmental exposure for each woman in the study, using data from GIS mapping (Brody et al. 2002; Silent Spring Institute 1998b).

Silent Spring's scientific work has led to the discovery of new chemical household exposures that may prove valuable in understanding breast cancer risk and to the development of innovative epidemiological methods that were motivated by activist questions. Unlike the Long Island Breast Cancer Study Project, which examined chemicals no longer in use, SSI examines both past and current exposures. The institute has been tightly conjoined with movement activism and organizations. Researchers have taught activists about the science of breast cancer research and have spoken at public forums and protests about the importance of an environmental causation hypothesis. Meanwhile, activists from SSI's Public Advisory Committee have worked closely with SSI researchers to form research.

Laywomen from the Massachusetts Breast Cancer Coalition advise the research process through a number of formalized and informal pathways. One researcher described some of these pathways:

> We have the Cape Cod coordinator on our staff who is herself an activist. We interact regularly with board members for Mass. Breast Cancer Coalition, with advocacy organizations that are engaged in public policy debates and we get a large number of calls directly from members of the public. We hold community information sessions, we go to other people's meetings and speak. We go to fundraisers like the Race for the Cure. I always go to and speak at the Boston Race.

These collaborations result in close relationships between researchers and activists and ultimately in the development of trust between the two groups. This has been cited as one of the most challenging and important goals for collaborative research. The most formal participation takes place through the institute's board of directors and Public Advisory Committee (PAC). In the PAC, members of MBCC and Silent Spring researchers meet to discuss various aspects of the project. The scientific team includes the institute's staff scientists and coinvestigators from Boston University, Harvard University, Tufts University, and other research groups. A nationally recognized group of scientists who constitute a scientific advisory committee also meets annually. The researchers also serve as educators as they train laywomen in how to understand their research proceedings and breast cancer science in general. Researchers also serve as activists by advocating an environmental causation hypothesis and supporting activists at public demonstrations. The institute supports events throughout the year that further the activist and researcher agenda by educating the public and drawing attention to the activities of Silent Spring. For example, the annual "Swim or Walk against the Tide" uses the walkathon strategy to raise money for the institute. On that day, information about the work of Silent Spring is provided to those who attend.

Silent Spring serves as a scientific resource for activists throughout the EBCM, not just those in Massachusetts. One Marin Breast Cancer Watch (MBCW) activist said, "When we have scientific questions, we just call Silent Spring!" This shows the role that Silent Spring has grown to encompass in the movement; much more than a research institute, it is a place where activists can go to consult with experts about a variety of scientific subjects. Researchers' assistance aids in making participation effective in other circumstances by giving activists the knowledge they need to provide input on projects and translate science to movement constituents effectively.

Silent Spring was the first collaboration with EBCM activists to take a functioning democratic form and is still the most effective. However, the full participation of activists in research and the environmental focus of the institute have faced a great deal of scientific and governmental challenge, unlike other activist groups that have adopted more mainstream projects in order to gain more support, like supporting women undergoing treatment, and scientists who have used less participatory approaches, like informal advising.

EBCM activists were able to shape scientific studies as the institute collected new data in order to answer the complex questions posed by activists. Women on the Cape wanted to know what they were being exposed to that could be the source of the heightened rates in their area. By sampling air, urine, dust, and many other substances in homes, researchers at Silent Spring found sixty-seven pollutants in homes that were generally ignored in breast cancer research (Rudel et al. 2003). Many of them had long been banned and were out of use. Discovery of their presence in women's homes revealed the need to study them as possibly linked to increases in breast cancer risk. Introducing new chemicals for study has the potential to have broad social and scientific repercussions. Scientists and activists involved in work at Silent Spring recognized the importance of their work in that sense. One researcher said, "We do try to show how this research has broad implications. It's not just about breast cancer." In a very similar tone, an activist who works with Silent Spring said:

> In other words, we're doing this research on Cape Cod, but our goal, and of course what's going to happen anyways inevitably, is whatever we find as a result of our research work here will benefit women everywhere. Not just women on Cape Cod, but women everywhere. This is the lab. We're doing the work here, but the work will affect and help women worldwide, really.

Collecting New Data, Constructing New Analyses

Like the EBCM, ADM activists needed to collect new data to bring their agendas to bear on government agencies. People who lived where dams were being planned or where there were high rates of breast cancer gave information to researchers about displacement, land values, or toxins to which they might have been exposed. This localized or embodied knowledge was often subsequently used in scientific analyses or technical reports. These became peer-reviewed studies or reports submitted to government agencies. Although these two types of documents have different forms and purposes, they are basically the same in development. By implementing lay perspectives in these documents, a two-way process ensued. Lay perspectives were legitimized and supported, and scientific understandings were altered.

One of the more successful CSAs in Brazil was in southeastern Minas Gerais, where many dams are either in planning or under construction. The first lay/expert collaboration in Minas was initiated in the mid-1990s at the Federal University of Viçosa in the southeastern part of the state. Anti-dam activism began there in 1995 when the Commissão Pastoral da Terra (Pastoral Land Commission, CPT), the Catholic Church, dam-affected people, and researchers from the Federal University of Viçosa came together to attempt to prevent further dam construction in the municipalities of Ponte Nova and Guaraciaba. The Department of Rural Economy instituted an extension project that involved a multidisciplinary group of professors and students, with the collaboration of church and NGO activists and community members.

One researcher who had studied the southern origins of the anti-dam movement had recently become a professor and had begun to see companies moving into the area to construct dams. This researcher began to contact professors in order to initiate a study of the situation. He was already involved in a rural education project that connected him with an activist who was studying to be a priest. As community members began to hear about the new extension project that would assist people affected by dams, they became involved, and the group of both local people and professors grew. Together, locals and researchers identified four main areas where the community needed technical assistance. State funding to support research related to the extension project was granted in 1997 to involve over ten students and four or five professors. At first, researchers served as activists and educators. Although researchers were not campaigning publicly, it was only through their initiative that the movement was stimulated.

Activists requested a copy of the environmental impact assessment from the state environmental agency, Fundacao Estadual do Meio Ambiente (FEAM). Researchers translated the EIA into more understandable language so that the community could be involved in analyzing it. Since, as in other communities, many members were illiterate and not very mathematically educated, researchers verbally summarized the EIA findings. This is similar to another case in Minas where researchers developed a mural to visually portray EIA results. The local people then provided their understanding of the local context so that together with researchers, they could provide a counterargument to the official EIA. For example, although official figures reported that the dam would potentially affect a total of 370 families, lay estimations brought that number to 972.

The collaboration between researchers at the University of Viçosa and local groups brought local, disempowered people together with church and other community groups. They met regularly to discuss the research and to exchange ideas about political action. The exchange of information and experience led to both an effective critique of superficial EIAs and previously lacking solidarity. This consolidated what became one of the most active anti-dam groups in the country. The researchers' acknowledgment that local knowledge of environmental impacts was greater than their own understandings demonstrates a form of legitimacy that had not previously been offered to communities. Between 1996 and 1998, the collaborative group was able to halt the construction of three dams: the Pilar, Cachoeira da Providência, and Cachoeira Grande projects. One researcher who worked in this area said:

> In terms of analyzing the strategy of the EIA plans with the community, it empowers the community politically and tech-

> nically to have their input in the public hearing to affect the outcome. . . . I consider that a really important part of our work in terms of working with the community as an education process, looking at these plans, not just to organize to come out in numbers and make a lot of noise to defend their interests, but also to discuss and have some input on technical issues.

An activist pointed out the difficulties with the process, but also how revealing it had been:

> There was a problem with language. . . . Another was the central nature of capitalism and the interests involved in the game. So this was a space with a lot of disputation including the vendors, companies, and the researchers. The process was interesting; it was for the first time a time to really discuss dams. It was very unknown. It was the only place to have this discussion about the fact that the things were concluded that the dams are not all as great as they are supposed to be.

Because of the grassroots nature of organizing and working with researchers, the process of discussing and critiquing the reports took time and a great deal of effort. Semiprofessionalized activists often played a role in facilitating exchange between the researchers and the less experienced members. The dam-affected people also often lived far away and did not have their own transportation to the local university. Although some of the protestors had unparalleled skills in making public speeches and protesting, they had less experience with the technical aspects of dam building and specifically shaping policy decisions. They were also intimidated by encounters with government officials. Therefore, researchers served as activists in order to mediate these challenges. Although not all of the research/extension group's collaborative dam resistance campaigns were successful and several dams they protested were built, in many cases where contested dams were built, activists and researchers were invited to advise how to design indemnification projects. Researchers agreed to take part because they felt that because of the decision to construct the dam, this was a very important chance to improve the lives of local people.

The citizen/science alliance was the most effective form of lay/expert collaboration in creating new scientific studies. Citizen/science alliances in southeastern Minas Gerais resulted in new understandings of the impacts of specific dams and an overall understanding that official environmental impact assessments were often inaccurate. By working together to assess these EIAs and add information that had been omitted, researchers and activists provided new bodies of scientific knowledge. The collaborations in Minas Gerais conducted these studies that were then submitted to government agencies.

In other cases of anti-dam activism, new knowledge about environmental and social outcomes of dams was also produced that could not be directly related to policy change. For instance, the collaboration between IPPUR and MAB generated a new database of information regarding environmental and social impacts of over one hundred large dams in Brazil, information that had previously been unavailable. Without this information, there is little ground for the movement to stand on and make claims about the negative impacts of dams.

Changing Participatory Norms

Most democratizing science movements aim to develop improved participatory structures. The ADM and EBCM faced similar obstacles to that goal, such as limited institutional parameters and norms that give experts legitimacy and marginalize lay perspectives. By engaging in lay/expert collaborations, researchers, laypeople, and some government officials learned the normative and structural constrictions on what they called a "deepened dialogue." As laypeople inserted their knowledge into research projects like citizen/science alliances, changes in science took place. They included broadening norms regulating research and participation, as well as changing environmental impact assessments and other research. One anti-dam activist said:

> In the process of environmental licensing, there is a public hearing. Affected people participate. But we have arrived at the conclusion that the public hearing is a grand theater. People come, participate, speak. But they don't have power to make real decisions. So we think to have effective power in the decision-making we need to have a deepened discussion and a study before the licensing takes place. I have no idea how in an environmental impact assessment the government in this region can arrive at the conclusion that "the best development project here is tourism." . . . A company contracts other companies to do the study and for the government, then they give it to us 40 days in advance! This is absurd! . . .

> We need an institution that can be neutral, it could be the state . . . that creates real participation.

Lay/expert collaborations were meant to improve local participation in the face of these superficial structures so that laypeople could participate in assessing environmental impacts through public hearings. One scientist in Minas Gerais who collaborated with the movement claimed that one of his goals in working collaboratively with community members was to increase their ability to have effective input:

> In terms of analyzing the strategy of the EIA plans with the community, [it] empowers the community politically and technically to have their input in the public hearing to affect the outcome. This is one of the interpretations of a strategy that worked in three cases from 1996 to 1998. If you look at the literature on empowerment, I consider that a really important part of our work in terms of working with the community, in terms of an education process, looking at these plans, not just to organize to come out in numbers and make a lot of noise to defend their interests, but also to discuss and have some input on technical issues.

The ADM pushed for participation in policy making, as well as in the EIA process. Possibly because of the institutionalized role of EIAs in the public hearing process, discourse about one seemed to overlap with the other. This intersection of scientific development and policy making is an important one. The EIA and the public hearing process are common in many countries and in the approval of many types of projects. Democratizing science movements' agenda to achieve regulatory reform functions both through scientific change and other methods external to it.

Like the institutional forms and norms of participation embedded in public hearings, breast cancer science and other health issues are tightly regulated by scientific norms about who should participate in science, what subjects should be studied, how that science should be performed, and how results should be interpreted. CSAs shifted these norms by demonstrating that laypeople should be involved in research, that environmental toxics should be an area of research, and that innovative methods can be more effective than traditional ones.

A primary step to achieving any changes in scientific norms was the change in attitudes of both scientists and activists that came about through lay/expert collaborations. There were two main attitude changes: activists' attitudes toward researchers and the scientific process, and scientists' perspectives on the role of, and respect toward, advocates involved in breast cancer science. Activists described drastic changes in their perception of the amount of information science could capture about environmental causation, the length of time necessary to conduct research, and the processes involved. As laywomen became more involved in science, their distrust decreased and they saw the difficulties that scientists faced. Their initial fear and anxiety about interacting with researchers developed into mutual respect and comfort with them: "The thing that I came away with that was most surprising was how much the scientists and the MDs have come to value the activist perspective on these panels and not only just putting a face on the statistics, but also that they appreciate that when you ask the questions, 'Why is this relevant? Who cares?' " One activist pointed out how she came to feel appreciated, and how her input made science more accountable: "These scientists used to work in a vacuum, not where it was happening. . . . The majority [now] definitely want the input of the community." This change in attitude and increased feeling of respect in scientific circles are in direct contrast to what scientization usually does to laypeople—creating a feeling of marginality and meaninglessness of their embodied knowledge. This process resulted in lay knowledge being acknowledged as useful and important.

Scientists also had some initial fear and prejudices, but mutual respect was developed. Prejudices on the part of scientists were largely based on stereotypes of activists as "hysterical women," while advocate preconceptions were founded on insecurity in relating to scientists. As the obstacles most frequently cited by both researchers and laypeople, these apprehensions and prejudices were some of the most serious obstacles to engaging in public involvement research. However, in the end, scientists often greatly appreciated both the input activists supplied and the activist efforts that funded their research projects. The more interaction that researchers had with laypeople, the more they developed sympathy, understanding, and respect. One researcher who had worked with laypeople for some time described her experience in a meeting where outside researchers and breast cancer advocates sat:

> I was talking about ways that scientists could become advocates in priority setting and another guy in my break out group said, "Well, I don't have time to be politically active."

> He literally said, "I'm in my lab working so many hours a day, I can't do that." In that room was a woman with a job, kids, a house, and chemotherapy. She had time. It's mind-blowing. I feel like saying, "Will you please get your act together here?" I feel like I have to have a role as a translator and to try to bring attention to the common goal.

Without the development of that trust and mutual respect, activists felt outnumbered and marginalized in group discussions. They therefore participated less and consequently had less influence on the outcome of the deliberation. But when they had experience with scientists and collaborating, activists often became so educated about methodology that they had specific recommendations about how science could be improved in the future. Therefore, the attitude changes on the part of both scientists and activists were critical to promoting a more equalized dialogue between the two groups and establishing communicative action that resulted in the democratization of knowledge. The development of trust between scientists and laypeople was one of the pivotal aspects that enabled lay concerns to be adopted into official expert discourse.

One of the fundamental parts of opening a discursive space between activists and scientists was the sharing of language. This difficulty is common to the two cases discussed in this book. Participants noted the different vocabularies that each constituency used as a barrier to sharing knowledge and opening collaboration. One researcher said:

> And I think that some advocates get totally turned off because it's difficult for a lot of people in science to which they have been trained to talk. And which they're comfortable talking. But that's also just a total human thing. That everybody is just most comfortable talking in the way that they're accustomed to talking. So it's very difficult when you're talking to somebody that doesn't necessarily have the same knowledge base that you have, whether it be bigger or smaller, to adjust your language to that.

Technical, scientific language was often difficult for activists to learn but gave them power and authority they had not previously possessed. By learning scientific language, activists could understand the scientific claims being made and therefore better contest them.

For dam-affected people participating in research, barriers between researchers and community members appeared to be an outcome of educational and language differences. This meant that there was an additional layer of activism, lay experts who had sufficient background to readily overcome the language and educational barriers. Often, they were from NGOs. One researcher said:

> There are misunderstandings between modern and local worlds. The problem is with different ways knowledge is expressed. . . . There is a huge gap between technical language and local language. We try to get the technical people to accept local knowledge.

This gap is demonstrated by an example of an activist from an NGO:

> This guy at this meeting last week on energy alternatives, he's from MAB in the south. He said, "You have to excuse me if I don't understand all the terms that you are using. I never got past the 3rd grade. The first school that I went to blew away. The second school I went to burned down and the third school I went to went underwater when they built the dam." . . . A professor prepared a PowerPoint presentation to present to the technical experts at the International Development Bank. We taught this guy how to do it. He said, "But I don't know what to say!" We said, "Just look at how they do it. They just put it up on the screen and they read it." So he did.

Grassroots activists or community members often learned how to turn their knowledge into what is considered technical information. Researchers, and sometimes activists from NGOs, helped with this process. NGO representatives usually had a different educational background from that of grassroots activists and therefore faced fewer obstacles in this process. One grassroots activist said:

> Even when we have very active militants of the northern NGOs, they come from the same world as the guys from the World Bank. Sometimes they come from the ground, the same university. They speak the same language.

Although this difference between grassroots and NGO representatives occasionally resulted in divisions, NGO skills were more often used to assist grassroots activists in engaging with political representatives and scientists. Crossing barriers of language was often the first step to working with experts.

Ultimately, although changes in science did not always have immediate impacts, they were a part of making longer-term sustainable change. This functioned fairly similarly for both movements, with lay involvement becoming increasingly accepted. When the ADM first became active, local communities were often not even notified that a dam was being built. The movement gradually inserted itself into these government institutions, largely through the lay/expert collaborations that they developed and through protest. Over time, the EBCM and breast cancer advocates became much more involved in multiple institutions of the NIH, advising research selection and programs. This enabled movement actors to critique policy making and suggest changes.

Changes in Topics and Methods of Study

As scientists and laypeople worked together, scientific methods and topics began to change. In particular, environmental factors were increasingly addressed, and interdisciplinary methods were employed. Research questions, the variables being studied, and methods shifted, resulting in new scientific outcomes and policy advances. Activists changed both whose knowledge was considered in research and the amount of resources directed toward new types of study. They argued that expert knowledge is limited in that the researchers themselves do not possess the understanding of exposures or facts regarding local land that activists have. One EBCM activist said:

> Power isn't only knowledge; it isn't only that I know how to run GIS statistical analysis. That's not all power is. Power is perspective; it is sensibility and I even think power is intuition too, even though they would probably totally disagree with that. That really goes against the principles of what science is but my new motto is science isn't a science, it's an art. . . . Maybe you need to use the scientific process but you need to broaden the approach, broaden what you would ask, think of a new angle. There needs to be a lot more flexibility around doing that than there has been. Part of it is bringing new ideas and ideas to the table that scientists may not think about.

In this way, the claims leveraged by activists in the EBCM are both normative and scientific.

By becoming highly educated about the science involved, activists critique the methods, topics, and outcomes of breast cancer research. For example, after learning the science of breast cancer research, activists have claimed that existing environmentally related breast cancer research repeatedly studies the same body of chemicals with the same methods, resulting in little useful knowledge about causation. The frequent focus on DDT, which was often sprayed in the United States in the 1950s, exemplifies this bandwagon approach to science that activists argue has led to discrediting environmental claims. Scientists rationalized the selection of these chemicals by explaining that they were easiest to study. Chemicals such as DDT and other pesticides that were sprayed consistently over many decades on the Cape and Long Island remain in the blood and are therefore easier to analyze. However, the focus on these exposures resulted in mixed findings and gave the impression that chemical exposures should not be linked to increases in breast cancer risk. EBCM activists make the highly scientific claim that DDT passes through the body to the degree that it is impossible to know the exposure level of a particular woman. Therefore, it is impossible to correlate exposure level with disease contraction.

The EBCM's ability to generate a whole new set of studies focusing on environmental causation is striking if one considers the general marginalization of environmental health science in governmental institutions, and breast cancer science examining environmental factors. The EBCM has garnered more than $180 million in research funding that is used specifically to study environmental causes of breast cancer. The movement was also able to institutionalize lay involvement in science. It helped create one major research funding mechanism, the Congressionally Directed Medical Research Programs, which now funds a variety of research with lay participation, has placed laypeople in advisory positions in a variety of other projects, and has been instrumental in the creation of NIEHS funding of breast cancer centers to study the environment with lay participation (McCormick et al. 2004). These are all major innovations if one considers the prior and persistent scientific belief that research should be visibly separated from social concerns.

The science generated by the EBCM was methodologically innovative and pioneering in its focus on new variables. The EBCM brought a diversity of environmental risks to the fore to be studied by top-notch researchers. At Silent Spring Institute, researchers created new epidemiological approaches that revealed a host of previously unstudied or understudied chemicals that women are exposed to in everyday life (Rudel et al. 2003). One innovation was a scientific "shopping trip" in which researchers collected samples of air and dust from stores and homes to be examined for endocrine-disrupting and carcinogenic chemicals. These new methodologies collected information about exposures that women have in multiple locations and over time. This focus on the experience of women in their day-to-day lives was quite different from most breast cancer research that examines preselected chemicals in the lab. Through these techniques, researchers were able to conclude that hormonally active agents are common and important indoor exposures that require further research, and to characterize chemicals previously unstudied (Rudel et al. 2001). Establishing the existence of these chemicals is the first step toward creating broader scientific study and pointing to the need to regulate them.

Advocate interests changed the scale of study and the variables considered. In the LIBCSP, the GIS map was originally meant to be created at the county level. Laypeople pointed out that there was too much environmental differentiation within counties, so there needed to be a smaller unit of analysis. Therefore, the GIS mapping was done by zip code. Community concern about particular exposures was also considered. For example, radiation exposure motivated researchers to include it in the GIS study despite the generally low levels scientists believed that residents experienced and the difficulty in measuring low-level radiation exposure (National Research Council 1972). The multidisciplinary approach that had to be used to answer the more complex questions driven by public involvement also made science more comprehensive and accurate.

Breast cancer advocates also began to shift the research paradigm away from a biomedical approach to one that dealt with environmental concerns and recognized the influence of politics on science. By connecting research to concerns of the affected population, involvement of laypeople changes the motivation for research and leads to reevaluation of canons of scientific objectivity. This shift need not result in a loss of scientific objectivity but rather can alter the topics being studied. This process is only possible through the opening of the scientific process in which science is made less rarified. One researcher explained how social movements play a role in this process:

> In general, what social movements do is definitely raise concern and raise knowledge about the disease that they advance

> the study of. Whenever that happens you're going to get the ability to look at the disease maybe from a broader range of perspectives. In fact, I was just telling someone the other day that scientists are just people just like anyone else. It's the very first thing. They are human beings first. And what human beings have a tendency to do is to go for fads. So there get to be fads in research. There get to be dogmas in research of what an acceptable area of research is. So one thing I find very helpful about having a diverse group of advocates is that it can sometimes help to . . . loosen up whatever the current dogma is and get people sort of unentrenched; get people out of whatever dogma trench they're in.

Although Brazilian researchers had a different initial commitment than did American scientists, their political perspectives also influenced why and how they approached research. One researcher explained why:

> Intellectuals here are very close to social movements. . . . This is because of the Brazilian relationship with Marxism. The history of Brazilian politics includes a strong influence of Marxism. Social scientists have a certain involvement with social battles; like ___ for example. They participate in these fights. The level varies. There was a level of participation during the military dictatorship. This involvement is often driven by an ideological and not an academic motivation.

In this sense, the participation of anti-dam activists in science was often brought forth because of a preexisting commitment on the part of scientists to serving the community. Although there is much research that does not take local perspectives into account, this background made researchers more open to anti-dam activists than they might have been otherwise. Therefore, the ideological changes around participation were easier to achieve in some ways. However, the credibility of researchers was still challenged. One researcher said, "[When you work with activists,] they think that you are biased, not scientific, biased. But they have to swallow it . . . they only accept it because I produce." Like the EBCM, anti-dam activists shifted a narrowly defined model of research into one that accounted for more factors—in their case, environmental and social impacts. They shifted the perspective on dams to internalize previously externalized costs. Accomplishing this often meant changing the methods of study to use more anthropological or sociological methods and collecting different data.

Environmental and social costs were included in many instances around the country, both retroactively and proactively. For example, new research regarding Tucuruí Dam in the eastern Amazon shows how the ADM vastly expanded conceptions of impacts. When Tucuruí was constructed in the 1970s, environmental and social costs were largely ignored. Local communities displaced by the dam were entirely marginalized by the process, to the degree that even the compensation packages created in the planning process were never awarded to them. Corrupt local officials kept the money for themselves. However, after the global anti-dam movement pressured the World Bank to initiate the World Commission on Dams and Tucuruí Dam was selected as one of the case studies, a much more comprehensive view of impacts and problematic processes was generated. The commission reported that its "framework for decision-making is based on five core values—equity, sustainability, efficiency, participatory decision-making and accountability" (2000, 3). These are mostly the values presented by dam-affected people. For example, in a letter of protest regarding the possible construction of a dam in central Brazil, MAB said that the company "has failed to respond adequately to repeated efforts by the affected populations and other civil society organizations (MAB 2002)," or, in other words, to be accountable to the interests of local people, one of the core ideas of the WCD.

Gathering these new data necessitated a focus on different variables and the use of new methodological tools. For example, researchers in Minas Gerais used anthropological methods and provided more detailed analyses because the EIA used engineering and physical science approaches. The team reported that the EIA "is a schematic proposal, assembled from an engineering manual without demonstrating what is included in the local situation. It does not analyze the changes due to local topography, the necessity of localizing construction, the suppression of native vegetation, or the destruction of land (Zhouri 2002)." These expanded scientific processes expanded reporting on local impacts and made science accountable to local communities.

Collaborations between activists and experts in Minas Gerais and other states have been very effective. The movement has been consistently engaged in lay/expert collaborations that have alerted community members to the new construction of a dam of which they were not informed and has taught people about the technical aspects of environmental impact assessments and energy policy. It has occasionally been able to stop dam construction. In some periods, changing science has been an important tactic of these activists.

Reframing Science

Engaging in and democratizing science benefit movement development and change science, often offering movements more credibility for their struggle. DSM engagement with science also enables the reframing of issues. Reframing science is similar to the process in which movements create "frames" (Snow and Benford 1992) by reinterpreting experiences and events of their constituents and reorienting them into an interpretive schema. This schema often connects activist experiences with a social justice perspective meant to initiate and sustain movements. Public discourse and political language, in particular, are often reframed by social movements (Steinberg 1998). DSMs show that science is also a fertile ground for framing in order to create normative and ideological change. The process of reframing is often both an outcome of existing lay/expert collaborations and a step toward new collaborative projects. It involves translating what movement leaders have learned in new participatory research projects to their constituents by making expert knowledge more accessible for the broader public and by politicizing it. This step also involves translating existing science into lay language for their constituents so that this new frame is accessible to the entire movement. Activists use it in campaigns, protests, and public education. By garnering broader public attention for a particular argument, the movement attempts to gain further support in the form of either research dollars or policy changes.

In reframing, movements use expert language and symbols to contest the justification for particular approaches to disease causation, political decisions, prejudices, and many other public and private discourses. This reframing process shifts the focus of science and provides a new value system and motivation for research. It also creates a new way in which the public can view the subject being contested. In this way, it is a critical step in social movement success. There are three main steps for activists in reframing: (1) learning about the details and technicalities of research findings, (2) consequently creating a new publicly accessible argument, and (3) finally attempting to change the discourse of government and scientific institutions.

The first step takes place in the CSA. The second occurs as movements begin to form their own interpretation of science. The last takes place as that new construction is publicized. Discursive change is central in reframing and in creating a successful social movement (Kolker 2004). The primary role of expert knowledge in creating discourses about development, the environment, health, and a broad variety of other topics makes shaping it fundamental to altering scientific and public discourse as well. Statements made to the media, testimony in Congress, or other verbal or visual representations made in public spaces reflect movement discourse. More specifically, discursive challenges posed by movement activists involve claiming rights to, and using, expert language in order to contest policy or create new definitions.

For democratizing science movements, one purpose of engaging in scientific debates is to translate scientific findings and technical reports to the public in a way that turns traditional conceptions upside down. This is a particularly difficult task because of the general perception of science and expertise as unquestionable. However, each of these movements uses strategies that simultaneously challenge public perception and work toward achieving movement goals. The EBCM makes very public claims in newspapers and educates the public through workshops and conferences. The ADM develops mottoes for protests and dam sit-ins where reframing is a stepping-stone to stopping construction. The EBCM and the ADM have learned the intricacies of research and scientific paradigms in order to broaden conceptions of who is responsible for social change. Activists have challenged certain scientific approaches and findings and have proposed alternatives that draw attention to government and corporate responsibility.

Activist organizations used their understanding of the breast cancer experience, as well as the science and the information they learned through lay/expert collaborations, to reframe the illness. In this way, the lay/expert collaborations in which organizations engaged served as a conduit for creating broader movement claims. For the EBCM, this process revolved around educating constituents about new findings or critiquing existing individualizing approaches that put responsibility for prevention on women. The ADM created a new paradigm of sustainability based on activist involvement in research, which was an alternative to the construction of dams.

Creating a New Paradigm by Translating Science

Many EBCM organizations commented publicly on studies and translated them into lay language for their membership in newsletters. For example, one Marin Breast Cancer Watch newsletter in 2002 reported on a diversity of breast cancer studies, including the increase in risk caused by burned meat, the scientific debate about hormone replacement therapy, and even research being conducted by other EBCM organizations like Silent Spring Institute. The Prevention Is the Cure website sponsored by the Huntington Breast Cancer Action Coalition (HBCAC) on Long Island states:

> We differ from other groups in seeking to cure disease by providing science-based information to the public, and promoting informed decision making, reduced exposure to environmental toxins, and community wellness. The science has spoken! Harmful environmental toxins are causing disease. Instead of waiting for definitive "proof" of the toxicity of millions of chemicals and products, we advocate a precautionary health model—one that says it is "better to be safe than sorry." (http://www.preventionisthecure.org)

HBCAC activists have participated in collaborations with scientists. This quote exemplifies how HBCAC uses science to create a new model of breast cancer prevention and causation both by focusing on the scientific data that support an environmental link to breast cancer and by invoking the precautionary principle that many environmental health researchers promote.

Breast Cancer Action is particularly adept at translating complicated science to lay populations. Barbara Brenner, the director of the organization, has written numerous op-eds in newspapers around the country and articles in its newsletter about the debate over using tamoxifen as prevention and the efficacy of mammography (Brenner 2003). The organization critiqued both the science involved and the approach to prevention being used. It pointed out that using drugs to prevent breast cancer is profitable for large pharmaceutical companies and that other forms of prevention would have fewer side effects for women. In addition, each newsletter offers definitions of breast cancer scientific terminology in order to help nonscientists navigate their experiences and reduce barriers between experts and laypeople. Through this process, BCA activists have used their scientific savvy to make a new public claim about causation and prevention.

The EBCM specifically reframes breast cancer as environmentally caused and in the process translates new science about environmental causes to the public. Activists also embed positive messages about lay participation. The following set of print media articles reflects the EBCM's approach:

"Pink Is the New Black: Does 'Shopping for the Cure' Cheapen the Reality of Breast Cancer?" *Salon.com (2005)*

"10 Ways to Reduce Your Exposure to Chemicals," *Consumer Reports on Health (2007)*

"Even the Dust Is Toxic in Homes, Scientists Say," *Los Angeles Times (2003)*

"When Citizens Meet Science: As Activists Push a Research Agenda, All of Society Usually Benefits," *Newsday (2002)*

"Expert Warns Marin to Look at Pesticides; Researcher Says Chemicals Could be Cancer Cause," *Marin Independent Journal (2002)*

An op-ed written by the director of Silent Spring Institute for the *New York Times* exemplifies the types of scientific reframing that occurs. She poignantly critiques the scientific defensiveness of manufacturers of personal products:

> In response to a Yale study showing an association between the older dark hair dyes and cancer (news article, Jan. 24), the vice president for science at the Cosmetics, Toiletry and Fragrance Association dismisses the research because it is observational rather than clinical. A clinical study would select a representative group of women, randomly assign half to use the dyes and half to abstain, and wait more than 20 years to see how many get cancer.
>
> If that is the industry's standard of proof, it should do that study before it markets its products.

Activists and the scientists with whom they work have challenged dominant corporate messages in forums as well publicized as the *New York Times,* the *Boston Globe, Prime Time Live,* and billboards in city centers. Public critiques such as these have supported political campaigning and scientific development and have ultimately led to several regulatory advances and changes in public opinion. Because policy making is based on expert knowledge, the EBCM found it necessary to use scientific resources to advance its claims. Its involvement with experts gave movement actors a level of public legitimacy that could be gained through no other avenues. When newspaper articles placed quotes of women with breast cancer alongside those of their scientific collaborators, their voices were not just the "hysterical women" activists are often perceived to be. They held a more powerful role to make claims about the state of science.

Making Sustainability Central

The tools of reframing are different for the ADM than for the EBCM. While the EBCM uses newsletters, billboards, and some public protest, the ADM conveys the importance of sustainability, environmental values, and the importance of local knowledge through public protest and contests of corporations. This changes the conception of dams as economic development to that of dams as environmental and social costs. However, like the EBCM, the ADM attempts to reframe dams by critiquing the scientific and economic justification for them and by emphasizing the needs of local communities. For example, the ADM wrote a letter to the Brazilian Development Bank on the part of the Apinajé indigenous people contesting a dam in the state of Tocantins:

> We want to say that the earth for us is Mother and Father, "it is who raises us and not the government," because it is from her that we get our game, our fish, our medicines, and where we collect our fruits, principally *babaçu,* which is very important for our survival, where we get the thatch to construct our homes, and our utensils such as the *cofo* (to transport objects and food), the jirau, ladders, and others, and it is from the *babaçu* fruit that we extract oil, *farinha* (flour) and charcoal. . . . We want development for all, respecting life above all else and not a development of destruction and death which only benefits a small portion of our society.

Another portion of this letter demonstrates how local understanding of and dependence on environmental resources is pivotal to arguments made by movement actors:

> As we have learned more about this dam, we saw that it will directly affect the Apinajé People, bringing various problems, since the forming of the lake will flood more than 1000 *alquieres* (about 2,400 hectares) of the best land within our

> reserve, flooding two villages (Riachinho and Mariazinha), causing a reduction in game, fruits, *babaçu* palms, and the forests where we plant, these being natural resources which are fundamental to the physical and cultural survival of our people (MAB 2001b).

This letter highlights not only the emphasis on local indigenous communities but also how impacts on them benefit a small group of builders. In this way, the ADM reframes dams that are generally portrayed as a national economic benefit that generates jobs and livelihoods as something that destroys ecological resources and exacerbates inequalities.

Anti-dam activists also reframe the costs of dams by graphically drawing attention to the cultural and agricultural resources that will be destroyed by construction. At several public hearings, local activists brought samples of their farm products and laid them in front of the podium along with the MAB flag to display the worth of their land. This visual representation is in jarring juxtaposition to the environmental impact assessments read by dam builders that claim that the lands to be inundated are infertile. These devices for reframing are broadcast on a larger level when the ADM protests in larger forums like Brasilia, where the movement has held a number of marches on March 14, the Day of Dam-Affected People.[3] In those cases, thousands of activists are bused in from all over the country. Most of them wear the white MAB t-shirt and carry related flags. Banners and chants say, "Water for life, not for death!" This simple slogan leverages religious overtones provided by the Catholic Church and compares the source of livelihood that rivers provide with the destruction of livelihood and habitat caused by dams. It also draws attention to the increases in infant mortality, poverty, and social fragmentation that result in the destruction of communities and individuals.

Conclusions

Citizen/science alliances have the capacity to change research findings, scientific tools, and participatory norms. MBCC and MAB activists were able to do so because they were equal collaborators and were involved from the beginning of each project. Their agenda

[3] March 14 became the designated day for anti-dam activities at the first international meeting of anti-dam activists held in Curitiba, Brazil, in 1997.

drove the collaboration and specifically shaped research questions. In some senses, scientific expertise and methods were at the service of community concerns. The typology used in this chapter recognizes that participation is not only about what nonexperts contribute to expert arenas but also about what experts offer to movements. In order to consider this latter type of contribution, the idea of participation must be broadened to examine movements themselves, not just the scientific studies in which they engage. This conception can be applied to other social movements that have engaged with experts. Many of them, such as the AIDS movement and the KSSP mentioned in Chapter 1, have found learning from experts to be critical to movement goals. Other activists have needed the credibility of scientists to be able to gain the attention and legitimacy necessary to address their concerns.

Adding these components to conceptions of participation changes the understanding of who a social movement actor is while acknowledging the differentiated social roles between activists and scientists. It also contributes to theories of participation that often ignore the social contexts from which participatory research emerges. It demonstrates that relationships between citizens and scientists do not end when they leave the lab or the boardroom, but that there are much broader social and scientific ramifications of relationships and work between experts and laypeople.

Collaborations also transform the imbalance between scientific and citizen concern that drives research. Scientists are generally more concerned with scientifically identifiable risks, while the public is more aware of ethical, social, and ecological risks (Dutton 1984), making public involvement necessary to keep those other concerns primary. This process is like the concept of "participatory democracy" that is a result of public participation (Dutton 1984). Dutton states that the rationale for democracy in science is the same as the rationale for democracy in political decision making, whose consequences include, according to Rousseau and Mill, the education of citizens, the facilitation of legitimate decisions, and human capital development essential to an effective democracy. Krimsky founds the rationale for including the public in science on similar grounds: "What was once visualized as the exclusive and legitimate responsibility of society's scientific elites is now being viewed along more pluralistic models of decision making . . . because values are embedded in the process from the very outset" (1984, 59).

In a postindustrial "knowledge society," power relations are based on the ownership and production of knowledge (Gaventa 1993).

By broadening this ownership, the mechanism of public involvement changes the power structure that determines research, as well as related power dynamics. It institutionalizes the equal importance of diverse types of knowledge and therefore challenges norms of scientific legitimacy. Ultimately, research on public involvement argues that this method is a tool that teaches laypeople about science and sensitizes researchers to the concerns of the public in an age of increasing public distrust of the government and science (Dickson 2004). Collaboration overcomes the specialization of roles inherent in a bureaucratized society that has the potential to disempower some while legitimizing others. It is also central to DSMs' alteration of public conception through reframing an issue or topic of study. Sometimes this process is literal, for example, when an organization uses specific research findings to make an argument about regulating a particular chemical, and at other times activists develop broad frames that encompass many findings. Although evaluating the success of this reframing process is difficult, discursive changes indicate an incorporation of movement discourse into other realms. This discursive adoption is often the first step toward making concrete changes.

Reframing science for the public is the first step in causing a ripple effect throughout movements and the broader public. It is in some ways the midway point between citizen/science alliances and public action. Without this process of reframing, DSMs would have little ground to stand on and therefore would be trapped into making a difference only in scientific realms. Because scientization functions across the arenas of science, society, and politics, reframing and accompanying discursive changes are central to realizing the democratization of science.

Although there are many forms through which lay/expert collaborations and democratizing science can work, there are also many others in which they do not. Participating in scientific construction is not a simple task, and creating institutions that engender democratic participation is difficult. The following chapter examines how movements get co-opted through participatory research. It both calls on and challenges theories of participation. When structures limit participation by providing only informal or superficial pathways, participation often results in co-optation.

6 Democratizing Science as a Mechanism of Co-optation

Covered in dust from desertlike lands, Celia[1] was born and grew up in the northern region of Minas Gerais. Since colonial times, Minas has held the promise of stone-colored riches born from the labor of workers subservient to mineral-hungry elites. Lack of water kept this wealth from reaching northern Minas, where poverty spreads across dry plains. Minerians, as they are termed by Brazilians, are often small and tanned and speak much faster than people from other regions. Celia is a unique Minerian, tall and olive-skinned because of Arabic parentage. The imposing nature of Celia's physical appearance is only augmented by the strength and eloquence of her words. Her speeches are delivered in classrooms and government offices as she works with the movement of dam-affected people.

Celia is in fact a unique combination of *atingido* and researcher. When a hydroelectric dam was built in the river adjacent to her home, Celia was temporarily uprooted. After earning her Ph.D. in sociology, Celia returned to her home state with an environmentalist husband, ready to fight a battle against dam building in Brazil. She recognized the power of corporations that were building dams and chose scientific study as a way to contest them. Celia has been visiting the poor, rural lands of potentially dammed people for several

[1] Celia is a pseudonym used to protect the actual individual.

years, conducting a research project with the locals. The project has already altered the course of dam building in Brazil. By pointing to the inadequacy of corporate assessments of environmental and social impacts, her studies led to governmental support of local people. Thus poor farmers have stood toe-to-toe with corporate engineers, demonstrating that their knowledge is deeper than any advanced degree.

The collaboration that Celia led was one of the most deliberative and scientifically effective in all Brazil. Her project stands up to the qualifications of deliberative theorists and scholars on participatory research. However, the governmental pathways through which this knowledge could get used are not sufficiently participatory. Therefore, simply changing the research was not enough to make policy change. In fact, despite the good intentions of those involved and the benefits to movement development and organizing, such participation in research can serve as a mechanism of co-optation. In this sense, democratizing science is a double-edged sword. It does not necessarily result in the scientific outcomes that activists imagined, and it also does not always serve the political outcome that they had hoped for.

As described in Chapter 1, activism can break down into co-optation during participation in research or in political institutions. In fact, in the modern, democratic state, science and politics are so closely intertwined that breakdown in one realm can lead to breakdown in the other. One of the most important findings of the two cases discussed in this book is that participatory research and deliberative democracy often cannot be separated from each other. Ineffectiveness in one realm causes ineffectiveness in the other. At the same time, effectiveness in one does not necessarily lead to effectiveness in the other.

The last chapter described how DSMs deal with one of the three aspects of scientization—the biased production of expert knowledge. This chapter relates how movements often are unable to address the entrenched nature of political dependence on expert knowledge. DSM inability to successfully democratize knowledge production or policy making is often not caused by outright denial on the part of experts or public officials. Rather, it is a more nuanced and complex process through which DSMs' interests are partially incorporated. On occasion this means that the movement is able to achieve partial success in changing science or policy. On other occasions, experts or policy makers are able to claim that they catered to movement needs, but in fact movement interests were perverted.

The two examples in this chapter represent scientific paradigms and government institutional functioning around the world. Participation in research is increasingly common and subsidized by large government institutions (McCormick et al. 2004). The Long Island Breast Cancer Study Project was the first such project instigated by the EBCM and possibly the most controversial. Part of the controversy revolved around its particular form of research engagement, which in fact resulted in movement co-optation. The environmental impact assessment and the public hearing process are also common to many national contexts (Glasson and Chadwick 2005). However, as one case in Brazil demonstrates, it is inherently flawed. The EIA and participatory research often result in the co-optation of populations because they appear to take citizen opinions into account, often only to dismiss them. These examples across cases demonstrate the interconnections of science and politics, and that in either realm DSMs can be defeated or co-opted.

Cooke and Kothari (2001) have asked if engagement in supposedly participatory development is "the new tyranny." This is a useful question that can also be applied to participatory research and the movements that produce it. Do these projects truly answer the needs of and resolve the conflicts raised by affected populations? Many authors have argued for the need to account for citizen perspectives both in policy and in conflicts over science and technology. However, the way in which citizen perspectives are handled determines whether this is possible, or whether participation is actually a sham.

Knowledge in Political Deliberation

Contesting science is generated not only by the exclusivity of the scientific process itself but also by the impacts that science has on the public through governmental policies. Movements aim to shape research policy, often with the intent to form new regulatory policies. Democratizing science movements may attempt to do both but be thwarted in one arena more than in the other. While participatory practices of research institutions may change research methods and trajectories, the government institutions that use findings to create policy may be less participatory. In that case, DSMs are often limited by these institutional forms.

Expert knowledge and lay perspectives are what constitute input into political institutions. Deliberation is the process through which those knowledge bases interact. It is a linchpin in resolving conflicts over science, research, and technology. Through deliberation, citizens or members of civil society are expected to exchange ideas, reflect on differences of opinion, and come to mutual agreement about policy recommendations. Community members engage in a learning process and have the opportunity to develop an agenda of common action. Based on a broad range of case studies, a number of moral and ethical rationales have been offered (Gutman and Thompson 2004) for this deliberation that can operate in many contexts, such as aid projects (Dryzek 2002) or citizen juries and citizen conferences (Pimbert and Wakeford 2001) that justify recommendations for particular types of deliberative practices (Gastil and Levein 2005; Meehan 1996).

The literature on deliberative democracy has often ignored the interaction that citizens have with experts and how democratic institutions manage these different kinds of input. Although deliberative democratic theorists are generally less concerned with knowledge itself than they are with the structures in which it is exchanged, they inherently deal with the conflict between experts and laypeople. As Niskanen (2001) argues, political decision making is often based on centralized knowledge of those in bureaucratic institutions that is detached from legitimate local realities. Like most deliberative democratic scholars, he argues that nonexpert perspectives should be included in technocratic systems. The key to improving policy formation is to bring "ordinary knowledge" into decentralized spheres of political decision making.

Fung and Wright (2002) similarly claim that implementing knowledge held by local actors in deliberative spaces makes policies more effective. This is true for two main reasons. Novel problems may be better solved by citizens who have direct experience with the problem. The intention is to shorten the decision-making chain and consequently increase accountability and effectiveness. In these spaces, instead of experts preempting popular participation, citizens provide active input, resulting in lay/expert "synergies." This is in direct contrast to top-down decision making that experts usually conduct. Therefore, disenfranchisement and distrust are replaced by empowered individuals who can apply their intimate contextual knowledge to a particular problem.

The rationale for participation can also be based on "epistemic proceduralism," or the theory that the cognitive nature of moral dilemmas faced by society should be recognized (Estlund 1997). Decision making should not be based on majority rule or correctness but on epistemic value. Therefore, different types of knowledge are recognized as legitimate. Deliberation involving different types of knowledge specifically has also been shown to improve deliberative practice.

Gambetta (1998) argues that transfer of information between different parties is an important outcome of deliberative processes. It provides an opportunity to update beliefs and opinions on the basis of greater resources.

These theories are particularly significant for understanding how and why DSMs introduce lay knowledge into research or political institutions. The following cases exemplify the limitations that DSM activists often face and therefore how their work falls short of democratizing knowledge. The most critical factor is often the lack of formalized ability for activists to influence either the research or the policy process.

Participation or Co-optation in the Long Island Breast Cancer Study Project

The Long Island Breast Cancer Study Project (LIBCSP)—the interdisciplinary research project pushed through Congress by some of the first EBCM activists on Long Island—was one of the initial collaborative projects instigated by the EBCM. From 1993, when the project was founded, until the late 1990s, researchers met privately with activists, and public hearings were held in communities. These meetings were meant to open science to community influence and feedback. Activist concerns were partially addressed in the funding mandate for several advisory boards to be put in place for the GIS study, the electromagnetic fields (EMF) exposure study, and two case-control studies. Variables for the GIS study were created on the basis of a series of town hall meetings held on Long Island. Hundreds of local residents gathered in churches to be informed by the primary investigator, Merilee Gammon, and other researchers that the project was taking place and to ask which environmental exposures should be included. In 1999, four community hearings were held in different locations on Long Island to ask community members what their exposures in those areas might have been. Researchers posed questions about what industries, agricultural practices, and polluting sources might have existed before reliable data began to be collected in the 1970s.

Representatives of the Long Island Breast Cancer Network also met with Merilee Gammon on a regular basis. They served on advisory boards for studies and to review proposals. An Ad Hoc Advisory Committee for the LIBCSP met annually. The fourteen members of this committee included four women with breast cancer who lived on Long Island. This group, the GIS oversight committee, and the community hearings were the main ways in which women with breast cancer were able to give their input to the development of research.

However, some institutional norms prevented realization of full participation. For example, because of the traditional process of peer review conducted exclusively by scientists, laypeople were not involved in the initial phases of research design. The public was consulted and its concerns were gathered through public meetings only after the project was planned. This was insufficient and in some ways harmful, as became clear only after the study was completed. In addition, activist decision-making power was informal in all these settings. As a result, many women were unhappy with the level of involvement allowed. One activist said:

> My complaint is that we were not involved nearly enough in the breast cancer study project early planning, or at least not in the case control study. NCI really had its back up about including advocates at an early enough stage to make a difference. A lot of these discomforts that a lot of us have about the LIBCSP has to do with that, was if it had been formulated differently we would have liked it better.

Advocates felt that even though they were consulted during the process of research, many of the things they wanted addressed were excluded. At times, the specific variables important to activists were ignored by researchers, which resulted in a reversion to traditional approaches and topics. For example, the chemicals studied had not been used for many years and had already been studied in several important research projects with equivocal results. Hunter and colleagues had provided ambiguous results in 1997 about the same chemicals that Gammon chose to study for the LIBCSP (Hunter et al. 1997). When the results were negative for the LIBCSP, activists pointed to the fact that they had requested that other chemicals be addressed, such as those in pesticides, plastics, and cosmetics. Scientists instead chose to study PCBs, DDT, and chlordane, whose use had been curtailed years ago. These chemicals were selected because they are easier to detect in the blood and body fat than many others. Because they are now banned, it was clearer that exposures preceded the manifestation of disease. But their lack of current use made it close to impossible to know either the time of exposure to the chemicals or the amount women had come into contact with.

Ultimately, although EBCM activists instigated the LIBCSP and were consulted on the research, they generally felt that their input was not considered sufficiently. This is a classic case of co-optation where participatory mechanisms are put in place that ultimately marginalize community concerns while creating credibility for the institutions in charge. This co-optation may have actually exacerbated the debate that was launched against the project when results were released in the summer of 2002 and garnered controversial national coverage. The media portrayed the results of the case-control study conducted by Gammon, Santella and colleagues (2002) as finding little to no support for a connection between breast cancer risk and chemicals used on Long Island, despite the finding that women with the top quartile of polycyclic aromatic hydrocarbons (PAHs) in their tissue had a 50 percent increase of breast cancer risk. This finding validated twelve previous studies. This increase in risk means that breast cancer could be prevented in a large number of women if point sources could be determined. A few national sources, in particular, *Newsday* and the *New York Times*, portrayed the study as problematic because of nonscientist involvement. *Newsday* claimed that "political pressures, inflated expectations and the competing demands of activists and scientists have conspired to undermine the [study]" (Fagin 2002b). In a similar vein, the same author reported that "even though some researchers and news reports warned of the immense complexity of the science involved, they were outshouted by groups willing to encourage such outsized expectations" (2002b).

The media also claimed that public participation is dangerous for science. A *New York Times* science writer, Gina Kolata, wrote, "The study's scientists, in the meantime, find themselves trying to appease two masters, other researchers and breast cancer activists" (2002). Activists had pushed Congress to fund the study, and local women had been vocal in their involvement in guiding the research. Media sources partially blamed activists for the study's lack of success. Scientists from well-known bodies who opposed this backlash were unable to get op-eds or even letters to the editor published, including James Huff, director of the National Toxicology Program, and Lorenzo Tomatis, former head of the International Agency for Research on Cancer.

The media used this representation to simultaneously argue that an end should be put to claims about environmental factors in breast cancer. Activists, however, argued that it was in fact the scientists' selection of chemicals to be studied that made the study less effective than it should have been. Further, they claimed that public participation remains a viable and significant component of environmental health research. In the end, both researchers and laypeople agreed that the adverse and at least somewhat inaccurate media portrayal of activist involvement could have ramifications for similar future ventures.

Soon after the LIBCSP findings were released, questions began to arise both about movement collaboration with researchers and whether further environmental breast cancer research should be pursued on Long Island. In 2005, these questions were answered in a congressional hearing held by state representative Tom DiNapoli and Congressman Steve Israel at Cold Spring Harbor Laboratory on Long Island. Several prominent scientists who had been investigating environmental causes of breast cancer and children's illnesses testified to the importance of providing further funding and support for the work. Political representatives also made affirmative statements. Although the director of NIEHS, Ken Olden, could not be present, he had prerecorded a video presentation that was played to the audience and further emphasized the points made by the testimonials. Federal funding was subsequently awarded for continued community-based research across the country to investigate environmental links to breast cancer.

Irapé: Critiquing Technical Information

The southeastern region of Brazil encompasses the states of São Paulo, Rio de Janeiro, Minas Gerais, and Espírito Santo (from south to north). This region is the most developed and industrialized in the country. The agricultural milieu is a mix of the small farmer-owned lands in the southern region and large landholder predominance in the northeast. The most powerful interests are frequently international or national corporate investors or landed elites. Minas Gerais is the largest of these southeastern states. It has historically been, and continues to be, one of the most intensely mined states in the country. Mining is often the motivation for dam building since it is an electrointensive industry. In the eighteenth century, cities in Minas were decorated with gold-laden colonial architecture that reflected the wealth extracted from mineral-rich soil. Today, mining and agroindustry, as well as generation of steel, chemicals, textiles, paper, and industrial equipment, produce millions of dollars in revenues for Brazilian and international companies (Agencia Nacional de Aguas [ANA] 2003). This has led to high levels of inorganic pollutants (e.g., copper and sulfate), largely in urban areas like Belo Horizonte, as well as in the lower São Francisco River basin.

There are two areas of activism in the state of Minas Gerais. A group called GESTA (Grupo de Estudo Ambientais) was initiated in the late 1990s to address the Irapé Dam on the Rio Jequitinhonha, in the northeastern region of Minas Gerais. The group responded when the state electrical company of Minas Gerais, Companhia Energetica de Minas Gerais (CEMIG), proposed construction. The low population density of the potentially inundated area provided the necessary justification for construction. However, the dam was planned to affect five thousand people in forty-seven poor, uneducated, and historically unorganized communities.

The research team, organized by a professor at the Federal University of Minas Gerais (UFMG), informed the community that a dam would be built and provided the details of the project. Together, the researchers and the community discussed the potential impacts of the dam. This professor was interested because she came from an area in southern Minas that had also been affected by a dam. Without the resources necessary to transport the community fourteen hours south to the capital where it could occupy governmental or corporate offices, the opportunity to stop the dam lay mainly in the local public hearing. The research team worked directly with one local leader who had reached the highest level of education by completing elementary school to organize local people and bring them into the hearing process. In the initial phases of researcher contact with the community, feelings of powerlessness regarding the prospective dam led to a potential for violence. However, because of the organizing efforts of this leader and the researchers, the hearing was conducted with vast community participation and no violence.

The effort was far from being solely a local struggle. The German nationality of one researcher enabled the group to draw international NGO support and consequently increased activist visibility and influence. His connection with the international organization FoodFirst Information and Action Network (FIAN) resulted in FoodFirst adopting Irapé as one of its projects. This organization launched a campaign in conjunction with GESTA. FIAN initiated an international letter campaign through which for the first time the Fundação Estadual de Meio Ambiente (FEAM), the state environmental body, received letters from all over the world about one of its projects. This was particularly important because of the international list of funders, the largest external supporter being South African Light. The total cost of the dam was projected at $1 billion.

As is routine with any development project, FEAM analyzed the environmental impact assessment of the proposed project that would be called the Aimores Dam. This analysis was furthered by a new environmental impact assessment conducted by the university research team and the community, which had pointed to many problems and found serious inconsistencies in the data presented (Zhouri 2002). The document declared, "From the perspective of GESTA, FEAM's questions practically suggest a new EIA that will justify or halt the licensing process" (2002, 4). The inaccuracies that the team aimed to correct included the number of families affected, actual versus hypothesized land values, and agricultural productivity of the land to be inundated. The EIA had been developed on the basis of footage of the ground below shot by planes flying overhead. GESTA had families collect information regarding what plants and trees were present in the area, the productivity of their land, and who would be affected. These new pieces of information were added to the report the team generated. One researcher said:

> We educate the community about what's in the EIA. We ask them what is wrong with it, about the environment. They say it better than anyone else. . . . They [the government officials] have the technical arguments but the local people frame it differently. The local people have a different way of expressing themselves, but there is technical knowledge buried in it.

After generating this new environmental impact assessment, the group presented it at the public hearing. Fifteen hundred community members attended. Many of them adopted a tactic that had been used by other anti-dam protestors at the Pilar Dam protest held in Ponte Nova in 1997 (Rothman 2008, 343). Community members brought vegetables and agricultural products that they had produced on their land to the hearing and displayed them in front of the speakers. These demonstrations helped support another argument the team leveraged regarding the cultural importance of the people in the area. One of the active groups involved in working with the researchers was a band of artisans who create brightly colored ceramic figurines. They are the last group of their kind, and this unique quality was used to demonstrate the importance of preserving the land and community. As the report claimed, the EIA "reveals a tendency to disqualify the material and cultural significance that the land has for these rural families, as well as for their economic and social activities" (Zhouri 2002, 10).

Ultimately, recommendations by FEAM's technical staff were superseded by the State Council on Environmental Policy (COPAM), which approved licensing for dam construction. The Assembly argued that activist fears that CEMIG would not provide resettlement were unfounded, and that the dam would provide development for the region. Meanwhile, CEMIG showed its untrustworthiness by launching Action no. 200.38.00.019793-0 to remove official recognition of the potentially impacted *quilombo* community and argued that people on this land should not receive indemnification because of its poor quality.

The case of Irapé demonstrates that even when new studies can be developed, achieving political change is difficult. Although FEAM acknowledged the improved validity of the new EIA and accounted for it in its decision making, COPAM overrode its decision and proceeded to approve construction. Collaborations with experts are critical to activism as local knowledge was pivotal in countering government and corporate claims. Experts also need local knowledge to prove the inaccuracy of corporate-generated EIAs. In this case, activists collaborated with researchers, who in turn contested government decisions. However, the weak and co-optable nature of state institutions showed that institutional form and strength of state agencies influence activists' abilities to attain their goals.

Mechanisms of Co-optation

Both the EBCM and ADM, as well as many other democratizing science movements, have floundered because they have been co-opted. Understanding the mechanisms through which this occurs can inform both how to create more participatory institutions and how movements can approach topics of contestation in order that they can actually be effective. Three main types of co-optation are exemplified by these cases: discursive co-optation, informal participation in political decision making, and informal participation in research.

Co-opting the Debate about "Environmental Causes"

Government institutions that have traditionally focused on lifestyle and personal responsibility have consequently used their support of the LIBCSP to shift the definition of environmental causes of breast cancer from human-made toxins to lifestyle factors that are not genetic. This undermines the work of the EBCM while allowing such institutions to proceed as though they were addressing the movement's concerns. This is a discursive form of co-optation that emerges from participatory co-optation.

After the completion of the LIBCSP, the National Cancer Institute, in conjunction with NIH and NIEHS, published a booklet on environmental causes of cancer (Department of Health and Human Services [DHHS] 2003). At first glance, the publication appears to be supportive of the EBCM agenda and of investigating a link between cancer and toxic exposures. It reads, "Exposure to a wide variety of natural and man-made substances in the environment accounts for at least two-thirds of all the causes of cancer in the United States." However, the following sentence reframes what environment means by stating, "These environmental factors include lifestyle choices like cigarette smoking, excessive alcohol consumption, poor diet, lack of exercise, excessive sunlight exposure, and sexual behavior that increases exposure to certain viruses" (2003, 1). This wording conflates the biomedical paradigm with the environmental one. This booklet is featured on the homepage of the LIBCSP. It therefore removes the legitimacy of an environmental argument gained by the EBCM in garnering funding for the LIBCSP and instead confuses the environmental definition with more typical approaches to breast cancer research that focus on personal responsibility and lifestyle factors.

The difference between the NIH conception of the environment and that of the EBCM is clearly articulated by comparing the booklet with documents from EBCM organizations. For example, the Huntington Breast Cancer Action Coalition, one of the primary groups that were involved in the LIBCSP, claims that "as we begin to understand more about chemicals, our environment and our health, we can choose to err on the side of caution. One such choice is to stop applying toxic pesticides on our property. . . . Toxic pesticides on lawns are like toxic drugs in our bodies. . . . In the long run they actually weaken the body" (http://www.hbcac.org/fednaturally.html download, 2007). The EBCM conception of environmental causes of breast cancer is oriented around toxic exposures like pesticides, radiation, contaminated drinking water, and waste sites. Pesticides were some of the exposures that concerned women on Long Island as contributing to breast cancer risk. However, only a few pesticides that are no longer used in the area were selected for study (Gammon et al. 2002a).

Informal Participation in Political Decision Making

One of the more common problems with participatory institutions is the lack of formalized decision-making power offered to laypeople. This was true for both movements. The public hearing is the main mechanism through which community participation in dam planning is allowed. It falls into the later part of the formal dam-licensing process. Since the enactment of Brazilian environmental legislation in 1986, the dam-building process consists of three licenses: (1) preview, (2) building, and (3) operating. The preview stage consists of the prospective company submitting an environmental impact assessment (EIA) to the government. Following the approval of the EIA, a building license is given. After the construction is approved, an operating license is provided. However, the only window for public participation in decision making is between the first and second phases.

An open public hearing can follow the EIA if the local community requests it within forty-five days of being notified of construction. The hearing involves the presentation of the official EIA by a consultant, statements by government officials, and testimonies by local communities or their representatives. When a river crosses state boundaries, IBAMA oversees the public hearing process. When river boundaries are within one state, the state regulatory agency oversees the public hearing. The hearing is one of the primary spaces for the participation of civil society in environmental policy making (Rezende 2007). However, this space is far from adequate because, as an IBAMA licensing official pointed out, "Public hearings provide a very limited basis of what you can do because they are so short term, four or five hours, and the public doesn't have very much information." The lack of formal decision-making power for local people means that decisions are made by environmental agencies. However, financial or political interests can override the result of the public hearing process. A lack of formal community decision-making power at the hearing also means that even if the community does participate, it does not necessarily have any influence.

Professional research consultants hired by prospective dam builders generally conduct the EIA. These engineers are not trained to assess social costs, and they use methods appropriate to measuring only certain outcomes. Research is often performed by flying in planes thousands of feet above the river to count existing forest acreage and observable numbers of settlements. Details are often missed by such reports. For example, one movement leader and church representative explained, "You go to look at EIAs in the computer and practically all of them are the same. We observe that an EIA of one location has the exact same photograph as an EIA in another location. One is practically a copy of another." Since the EIA is the one point at which environmental concerns about dam building can be considered in the planning process, it is critical that these reports be accurate. Dam-building companies have a conflict of interest when they are conducting these studies, as do the consultants they pay. Therefore, IBAMA and its affiliates are the closest to an objective observer. The studies conducted for EIAs usually involve topographic, hydrological, geological, and environmental data (Fearnside 2005). The national governmental guide to inventorying the hydroelectric potential of river basins uses sources like the annual Brazilian Demographic Statistics (Anuário Estatístico do Brasil), state governmental information, and scientific publications to assess environmental and social impacts. These sources are broadly quantitative and provide little information about local context.

Similar to the supposition of many theorists that deeper democratic practice must devolve power to local groups, this movement pushes for the actual involvement of laypeople in the construction of EIAs that are required by the government and are presented at public hearings. This would result in a more accurate assessment of potential environmental impacts and therefore better account for the costs of dam building. These problems make the public hearing process only the first step toward building a forum where different perspectives can be shared and mediated. Because of the lack of avenues for participation, local groups coalesce to assert their opinions and gain rights. This results in movement formation if community interests are not met.

Informal Participation in Research

Like nonformalized participation in political institutions, lay participation in research is often informal and results in feelings of disempowerment on the part of activists after engagement in science. The CDMRP that partially funded the research on Long Island reflects these issues. The Breast Cancer Research Program works under this heading. The BCRP is "a partnership of consumer advocates, clinicians and scientists collaborating with the DoD to identify gaps in research design, new mechanisms for supporting research, and guide the funding process (BCRP 2003)." In order to do that, the program claims to have broad representation, equal treatment, orientation for participants, education about each role, and evaluation of the process, resulting in a consumer benefit score. In order to achieve that, program documents explain that consumers are nominated by an organization, and selection is based on rank of screening scores. One mentor and one novice are on each board, and there are always at least two consumers to a board. The CDMRP provides some of the largest federal funding mechanisms for breast cancer research and gave the largest amount of DOD funding for the LIBCSP.

Although it appears that advocate involvement is well instituted, there have been major critiques of the program (McCormick, Brown, and Zavestoski 2003), including a prescribed number and type of advocates. The number of members of the total panel means that advocates are vastly underrepresented, and their opinions can easily be outweighed by other groups. Although advocates are involved in the group discussions with experts about how a proposal should be judged, the formal role that advocates play is unclear in all these settings. Many advocates have felt intimidated by working with scientists and high-level officials. In addition, the advocates involved in the highest-level boards are often the same year after year, meaning that only a few organizations and opinions are represented. Although incorporation of activists appears to signify coalescing agendas between the breast cancer movement and NCI, DOD, and others, that is not necessarily the case. Activists may simply be taking the opportunities available to them at the governmental level. Meanwhile, these federal-level programs have grown over the past ten years, in particular, and claim that advocate involvement is fundamental to decision making.

This model of engagement has led to funding for research projects that do not relate to movement concerns, despite the fact that the EBCM instigated the growth of these research programs (Davis 2002). Although the CDMRP now funds research for several disease areas, breast cancer continues to receive more than twice as much as any other disease. Of all 168 epidemiological studies (which are only a subset of all the studies, which also include tissue studies and others), 14 address environmental components (CDMRP 2003). This 8 percent of epidemiological studies is an even smaller proportion of the total number of research projects supported by the program.

Conclusions

The outcomes of DSMs are not always what activists had planned or hoped. There may be insurmountable obstacles to egalitarian participation, and therefore the process of democratizing science can fail. Even when deliberation in research is effective, engaging in science may actually prove to be a waste of time when findings are ignored in policy making. While DSMs have been shifting the epistemological landscape, they have also consistently faced political institutional and paradigmatic constrictions related to citizen participation or scientific approach. These parameters leverage power and control for participants while marginalizing the concerns of other social actors. DSMs attempt to reveal these constrictions. Democratizing science can actually be a form of movement co-optation when either participatory research or participatory political institutions do not formally involve advocates. Specifically, this means that they do not have the capacity to vote or exercise any influence on decision making.

Another set of problems that leads to co-optation is lack of education about expert language or lack of knowledge on the part of advocates. When this happens, activists may be present, but their participation is limited. This can be remedied by researchers acting as educators or by institutionalized programs like ProjectLEAD, which trains activists to understand the science of breast cancer and therefore enables them to be involved in research. This project was initiated by and is currently hosted by the National Breast Cancer Coalition and involves a variety of breast cancer organizations (Braun 2003). The training is a four-day science education course that takes place four times a year. During the course, advocates learn the basic biology, toxicology, and epidemiology involved in complex scientific questions. ProjectLEAD, however, is focused on preparing women to be reviewers for DOD projects and includes little training in environmental research.

In both research in which the EBCM engages and the EIA construction and public hearing process for the ADM, there is a lack of participation that leads to movement co-optation. Both movements have expended a great deal of time in both types of activities. These are the few pathways through which they are invited by officials and experts to make change. Therefore, they have little choice but to participate. At the same time, by participating they are in some ways undermining their own agendas. Their efforts are often frustrated. Whether or not participatory projects actually help citizens or meet movement demands depends on the conditions under which the project is implemented. Co-optation may not be intentional, but it still raises questions regarding how to develop institutions that actually work for democracy.

7 Long-Term Struggles and Uncertain Futures

What does the future hold for democratizing science movements? Are we headed in a direction where science can serve as a pathway for democracy, or will it continue often to limit citizen influence? A plethora of the most controversial recent social debates have begged this question—from those over end-of-life decisions that the Terri Schiavo case raised (Koch 2005) to debates about intellectual property rights or bioprospecting in developing nations (Dorsey 2004), challenges to complex and value-laden science are being posed by nonexperts. Movements are forming in response, and policies are being made. This chapter examines the most recent and some of the most controversial projects instigated by the EBCM and the ADM. The purpose here is to look at the long-term struggles that each of these movements is now engaged in and attempt to assess how they and other DSMs remake themselves under changing global and scientific conditions.

The cases in this chapter——EBCM organizing in California that led to Breast Cancer and Environment Research Centers funded by NIEHS, and defeat and development of Belo Monte and Rio Madeira dams—indicate both what these movements may be able to achieve in the future and how they continue to raise concerns that go unresolved. They also beg the broader question of what role the lay public should play in charged decisions about science, research, and technology. Since they are both the outcomes of years of struggle, they raise and begin to answer several important questions. What happens as DSMs grow and become more established? Do they move from engaging in science to other activities? Or do they grow so scientifically engaged that the tactic is critical to organizing? Are they able to achieve more or less policy change?

Although the answers to these questions will not be resolved in this book, some recent findings point to the critical nature of these questions. Researchers are discovering that the public sees that scientists are influenced by economic interests and need to take new approaches (Luján and Todt 2007). Scholars also claim that debates over the impacts of new technologies lead to greater interrelationship between the public and scientists rather than more disenfranchisement (Stilgoe 2007). However, policies are still in place that discourage real citizen involvement in science (Schibeci and Harwood 2007). These concerns are present around the world, in Australia, Spain, the United Kingdom, and many other countries. Controversies relate to many different types of scientific and social developments. The two following examples tell us something about how engagement or disenfranchisement plays out, and where we will go from here. They raise questions of globalization, the trajectory of managing new biotechnologies, and whether movements will lead to more or less precaution.

Belo Monte: Shaping Science, Changing Policy

In January 1989, one of the few truly international anti-dam protests in Brazil took place to counter the proposal of a large hydroelectric dam plan in the eastern Amazon. Irish rock star Bono, international media attention, and powerful NGOs appeared. Later shown in the international press and several films, a Kayapo woman brandished machetes at an official (Fisher 1994). This protest put at least a temporary end to construction plans. Because these plans were coordinated by the most authoritarian state electric energy sector, Eletronorte, movement abilities to halt construction of Belo Monte on the sparsely inhabited Xingu River in Amazonas are striking. If constructed, the dam would be massive, with three times the generating capacity of Tucuruí; its construction would cost a projected $11 billion; and it would create a reservoir of 400 square kilometers (Pinto 2002). The majority of the energy was meant to be transmitted 3,300 kilometers away to more urbanized centers and industry in the center and south of the country.

Following the rejection of plans for Belo Monte, government planners went back to the drawing board and created a second model

TABLE 7.1 EARLY VERSUS LATER MODEL OF BELO MONTE DAM

	Before	After
Installed potential (MW)	11.025	11.182
Area of reservoir (km^2)	1.225	400
Soil excavation (mil m^3)	10.781	144.622
Rock excavation (mil m^3)	14.056	51.955
Soil landfill (mil m^3)	57.168	39.567
Rock construction (mil m^3)	16.765	16.957
Concrete (mil m^3)	3.494	3.841

that they called an engineering marvel. Table 7.1 portrays the differences between the first and second models. The left-hand column represents numbers from the 1989 model, and the right-hand column characterizes the new model developed in 2002. This model was technically far more complicated and innovative. Rather than a dam that would create a large reservoir, the new model was a run-through dam that depended on the river's current to generate energy. This meant that the reservoir would be much smaller, but that the depth of soil extraction would be much greater. The new dam plan published in 2002 showed major improvements such as a decrease in reservoir size and the rerouting of waterways.

This is not the first time a dam model has been marveled at or anticipated for generating minimal effects. Many other seemingly miraculous dams around the world ultimately failed, like Balbina in the western Amazon, which flooded massive amounts of land and generated very little energy. When the new Belo Monte plan was introduced, this common point of reference led to a general sense of suspicion on the part of protestors. Activists believed that not just one dam would be constructed, as Eletronorte claimed was the plan, but that after the first one was built, two to four others would follow. They based this perspective on their knowledge of the flat topography of the Xingu River basin, which necessitates the artificial construction of a stronger current; drastic seasonal variation in waterfall, which results in the need for a large reservoir to guarantee year-round water flow; and finally, past building practices where one dam was promised but many more were built. Although they were initially informed otherwise, activists' suspicions were correct. The Belo Monte plan included at least two dams, Belo Monte Dam itself, which would not create a massive reservoir, and Altamira Dam, to be built upstream to regulate flow (Fearnside 2006).

By the time Belo Monte was proposed for a second time in 2003, the international strength behind the anti-dam movement in the north had diminished, and expert collaboration with activists became more important in contesting the new plans. By joining local perspectives to an expert critique, these activists were able to temporarily shift governmental planning. Glenn Switkes from the International Rivers Network office in São Paulo published a critique of Belo Monte that demonstrated how it would displace indigenous people that the government claimed to protect (Switkes 2002). New collaborations were the basis of critical meetings such as one cohosted by the Development Movement of the Trans-Amazon and Xingu River (MDTX) where hundreds of indigenous people were in attendance. The increased level of expert participation filled in as support. In this way, activists adapted their methods to counter the more technically oriented governmental argument. Its engineering problems made the dam especially vulnerable to contest. An alliance between the movement and the public minister (ministerio publico, or the relative equivalent of the public defender in the United States) was also important. When he learned that Eletronorte had illegally attempted to get approval for construction from the state agency despite its need to actually gain approval from the federal-level IBAMA, he began to support the movement. As has been the case for many other dams, without activists drawing attention to these problems and experts or officials collaborating with them, the dam would likely have been built.

Several key activists played a role in the Belo Monte struggle. Take Monica,[1] for example. Monica traveled constantly throughout the Amazon and across the country to overcome the invisible divide between the forest's edge and the rest of Brazil. Monica was a leader in the MDTX, which was the largest activist network in the north. It was made up of 113 organizations that call for community participation in dam planning and that want justice for the murders of five of their members who were protesting construction. As is implied by its name, this network addressed the entire Amazionian region and especially the Xingu River that runs through it. Rubber tappers, indigenous people, fishermen organizations, and a variety of other groups

[1] Names have been changed to protect anonymity.

coalesce through this group. Primarily poor and geographically fragmented, these locals have had little political clout in planning.

The MDTX was initiated in the 1980s and addressed a variety of issues related to creating sustainable development. Historically, the MDTX worked in partnership with Forum Carajas, Living Rivers, and the Catholic Church protesting the construction of Belo Monte. The current senator of the state of Pará, two federal deputies, and a state deputy, a former activist, all worked closely with this organization. The activism of the MDTX and other local organizations has largely been focused on existing or potential dam sites. The large distances between sites have caused many difficulties in coalescing around one agenda. Interviewees also reported that differing agendas caused divisions in activist groups. For example, while some groups were more focused on large-scale development projects, others worked on local fishing rights, and others dealt primarily with indigenous rights.

Despite the movement's persistence and success at Belo Monte, the dam was reproposed in June 2005, and IBAMA speedily approved it. The EIA had been contracted out to an independent research institute, Instituto Nacional de Pesquiss da Amazonia, INPA, but the CNEC (the national consortium of engineers) had editorial control over the document (Fearnside 2006). The EIA was evaluated by state agencies and deemed sufficient. Questions were then raised about the real viability of and motivation for the project. Much of the energy would be used to process aluminum for Chinese companies (Fearnside 2006), and government representatives stood to profit from their financial investment. This contestation continued. Whether Belo Monte will be accepted and construction will begin is still an open question debated by activists and researchers alike.

Belo Monte is one of the most contentious dams in Brazilian history (Carvalho 1999). The twenty-year-long contest over Belo Monte and the movement's role in changing the shape of the structure's plan reflect its controversiality (Fearnside 2006; Turner 1992). Struggles over Belo Monte also reveal the importance of scientific contestation in changing policy. Since the movement in that area is predominantly made up of uneducated, marginalized indigenous people and fishermen, these researchers provided important legitimacy by speaking passionately at large protests and testifying in front of government agencies. More general conclusions from this case indicate that the movement can change the shape of future dam planning but will not be able to stop construction. More than any other, this case demonstrates the power of the anti-dam movement in changing policy and the limitations of that ability.

Expertise in Amazonia

Both indigenous knowledge and expert knowledge from researchers at the University of São Paulo (USP) have been critical to organizing campaigns against Belo Monte (Conklin 2002). Laypeople and experts worked in tandem with one another as researchers offered a critique of the dam that was supported by local knowledge of the area. Together, they held public meetings that hundreds of local indigenous people attended in order to learn about proposed plans that were otherwise unpublicized to them. At one meeting in Altamira, the city where most Belo Monte activism and planning was conducted, a panel of MDTX representatives and São Paulo researchers explained where the dam would be built and who would be affected.

The researchers involved in this collaboration also presented information to government bodies encompassing both their own perspectives on energy policy and local knowledge about problems associated with dams. They have drawn attention to alternative methods of energy production, such as wind, solar, and biomass, and have stimulated interest in these possibilities around the country. Two of the pivotal symbolic and practical pieces used in this process were books published by these researchers. Entitled *Brazil 2002: The Sustainability That We Want* and *Alternative Sources of Energy* (2002) and *Sustainable Energy in Brazil* (2000), these books offer exact calculations of alternative energy costs and a critique of a hydro-based energy model. Activists in both the south and the north use these books to support their lay perspectives.

Amazonian activists and their allies have used all types of collaborative forms. Researchers serve as educators by advising movement leaders on political practice, technical aspects of dam building, dam impacts, and renewable energy sources. USP and MDTX worked together to analyze this newer Belo Monte model and develop a technical critique. Through their CSA, locals inform researchers about the impacts they anticipate if Belo Monte is constructed. For example, experts technically codified local concerns regarding the size of the reservoir, which was in question because of the local flat topography and rainfall patterns. Local people knew that the rainfall patterns in the area would mean that for eight months of the year there would not be enough water to fill a reservoir. Activists estimated that the reservoir area would be larger than the government was projecting. By working together, these groups have created a new public discourse about Amazonian dams and the importance of local perspectives in their planning. But the most important collaboration between USP and the MDTX was the exchange of information and knowledge that led to new, more legitimate critiques of dam plans.

It is important to note that this case is special in the involvement of researchers from USP who are some of the most respected scientists in the country. Often, the legitimacy of researchers is challenged when they engage in lay/expert collaborations. USP researchers did not face this dilemma because they were already widely respected. This renown helped them engage in such a controversial struggle with success.

What the Anti-dam Movement Can Do Today

Contestation against Belo Monte reflects how the movement has developed over time, and how changing context has limited or enabled the movement. In particular, it highlights the growth of construction and activism in the northern region of Brazil and the ways in which these new developments may play out. In that area and others, the movement has long challenged the state, as well as the corporations with which it is affiliated, and has conducted public protest demonstrations, marches, and occupations of dam sites and public buildings. In many cases across the country, as well as in the Amazon, it has been able to achieve policy change. This process reflects the broader institutionalization of the Brazilian environmental movement (Viola 1997). At least, in many cases when a dam has been constructed, accompanying ecological parks and remediation have been mandated as a part of construction. For example, Eletronorte committed to the Parakana Program to improve the health of Parakana communities that were affected by Tucuruí Dam (Martins and Menezes 1994). Many electrical agencies now use these projects as promotional materials for their ventures, claiming that they represent environmental protection and awareness.

Since President Luiz Inacio Lula da Silva took office in 2003, the movement has been able to shape some of the most comprehensive policy changes at the same time that it has faced some of its greatest defeats. A new model of the energy sector was developed during President Lula's first term. Although it is difficult to trace the exact effects of the movement on this new energy model, main points of the movement's agenda have been incorporated into it, which is a significant achievement. The model included developments such as an active government role in planning dam building that might curtail the power of private interests to construct dams, and the conduct of environmental impact studies before a potential site is offered to private interests.

Although the World Bank's independent investigation of large dams, the World Commission on Dams, was not a government institution, its influence can also be seen in the new energy model. Its recommendations reflected some of the anti-dam interests and global norms about participation and environmental protection. Many government officials are aware of the WCD final report, and the Ministry of Mines and Energy issued an official letter that supported its outcomes. The WCD and the World Bank, which initiated the WCD in 1997, have been important mechanisms for policy making. One government official said, "International agencies have a lot of influence, like the WCD that is an international organ. It has influence here, putting ideas in the process." The World Bank has funded several dam projects in Brazil, as well as around the world. Despite the fact that proportionally it funds a much lower number of dams than other financial institutions, its global clout gives its practices important weight that others do not possess.

The new model was also based on a historical trend of increased awareness of social and environmental impacts of dam construction. For example, some of the new model's details may have been influenced by a meeting called "A New Model for the Energy Sector Is Possible . . . and Necessary" that was held by anti-dam organizations, the National Federation of Urban Services and Electrical Workers Unions (FNU/CUT), and several congressional representatives on November 6, 2001. Those civil society actors and social movement organizations began to reconceptualize what a more democratic electric sector would involve. One government official said, "There exists a consciousness in the electric sector now that is fundamental. You can't build a dam the way you used to here. It doesn't work anymore to build a dam the way you used to, with people coming in, building a dam and taking the money away all the time." Instead, costs previously considered external must be internalized in the planning process, and this changes the structure of dam planning, as well as when and where a dam is built. These changes represent great potential for citizens to be involved in dam-building policy; however, whether accountability increases depends on policy implementation.

Another Dam, Another Energy Model

Although this new energy model is potentially a serious breakthrough for movement actors, other more recent changes in the structure of IBAMA have undermined these advances. At the same time the new energy model was built and the second model of Belo Monte was defeated, a series of two dams on the Rio Madeira was proposed for the western Amazon. These dams were also developed as a part of the Integration of South American Infrastructure (IIRSA) project meant to advance development of Brazil, Bolivia, and Peru through 335 development projects, including hydrological development (van Djick and den Haak 2007). This dam system would be a part of interconnecting transboundary rivers to enable exportation of crops like soy to Europe and China.

The Rio Madeira projects entered the planning landscape initially in the mid-1990s as part of the federal-level plan called Avanca Brasil. Avanca Brasil was a much larger plan created by President Fernando Henrique Cardoso in order to organize development throughout the entire country (Laurance et al. 2001. The Rio Madeira complex involved two dams, one close to the city of Porto Velho and the other farther upstream, called Santo Antonio and Jirau, respectively. The Rio Madeira is not officially a part of the Amazon, because the flora and fauna in that area are slightly different from Amazonian rain forest. However, damming the rivers would affect the Amazonian forest in a number of ways, including road building, flooding, and deforestation through new landholding patterns. Figures relating to displacement and ecological damages are disputed, with estimations of displacement around three thousand people, six hundred species of fish, and seven hundred bird species (Cevallos 2006).

As President Cardoso left office and President Lula formed his administration, the Rio Madeira projects became a much greater part of the energy sector's consciousness, largely because of initiative taken by two institutions, FURNAS and Odebrecht, to study the viability of constructing dams in that area. Viability studies for the Rio Madeira dams began around the same time that the second model of Belo Monte was being developed in 2001 and 2002. When Belo Monte was again defeated in 2003, the Ministry of Mines and Energy proposed a new focus on Rio Madeira rather than Belo Monte. Intensified study of this site came from additional sources of political pressure. In 2001, under the administration of President Cardoso, Brazil faced an energy crisis that resulted in power loss and forced conservation measures (Szklo et al. 2003). President Lula campaigned on a platform of ameliorating this problem. His first term, however, was fraught with a mission to allay fears about inflation and economic stability, and he was only able to address this concern by initiating the restructuring of the energy model. Therefore, projected energy needs had not yet been satisfied, and the Rio Madeira complex was cited as a solution.

The motivation to pass new dam projects was accompanied by the development of new institutional structures that led to movement constrictions. Processes for formal licensing of dams were shifted to include an increase in the accountability of the already-stressed environmental agency, IBAMA. In order that each potential construction site be secured as much as possible before offering it to a company or consortium of companies, the overseeing IBAMA official who approved the site would be held personally accountable for it. If a site was ultimately deemed unusable, even after IBAMA approval, that official could be sent to jail without parole. This stipulation slowed all approval processes as IBAMA workers became afraid of facing criminal charges. Controversy over the stagnation of approval for dam sites reached its peak in early 2007 when President Lula called for a restructuring of IBAMA that would siphon licensing away from the larger IBAMA agency into a special presidential caucus officially a part of the military. IBAMA workers meanwhile went on strike, refusing to work under the existing conditions. They were ordered back to work by the federal judiciary body, which argued that IBAMA was hindering the economic progress of the country. As a result of this tumultuous process of passing new dams, IBAMA was restructured so that a new agency would conduct viability studies, consequently facilitating the passage of dams (Reuters 2007).

Before the proposal of the Rio Madeira dams, there was little social movement activity in the western Amazon. MAB did not have a presence there but was focused on work in the rest of the country. A small number of indigenous groups, international NGOs, and local environmental groups began to protest. Local groups included the Counsel of Indigenous Missionaries (Conselho Indigenista Missionário, CIMI), Canidé, River Earth (Rio Terra), Brazilian Network (Rede Brasil), and the Organization of Fishermen in Rondônia (Organização dos Seringueiros de Rondônia, OSR). International groups included the International Rivers Network, Greenpeace, and Amazon Watch. National Brazilian organizations such as Rios Vivos, the Movement of Landless People (Movimentos dos Trabalhadores Rurais sem Terra, MST), and the Movement of Small Farmers (Movimentos dos Pequenos Agricultores, MPA) were also involved. The Commission of People Impacted by Dams in the Amazon (Comissão dos Atingidos por Barragens na Amazonia, CABA) had been formed in response to Belo Monte and the proposal for other dams in the Amazon, and it moved into the Porto Velho area as the struggle against the Rio Madeira complex formed.

As the Rio Madeira complex quickly passed through planning stages necessary to license the projects, activists argued that they had not been given sufficient ability to engage in discussion. They referred to Law 10257/01, which guarantees that indigenous communities affected by development projects will have access to information about it and the ability to engage in discussion regarding its planning. They convened a meeting in Porto Velho supported by Amazon Watch and coordinated with local and transnational organizations like the IRN, the Pastoral Land Commission, and Friends of the Earth Brazil in May 2006. Entitled "Public Debate and Popular Capacity Building Regarding Environmental Licensing," this meeting presented the environmental impact assessment and allowed local groups and NGOs to speak publicly about their problems with it. Local populations argued that indigenous displacement had not been sufficiently accounted for. Polis, a Brazilian NGO, claimed, "The population needs to have access to the information about the project in order to monitor and evaluate its viability for local and regional development" (Franco 2006).

Consequently, international environmental NGOs (IENGOs) and Brazilian experts began to generate new studies and critiques of existing studies, resulting in several new reports. One of them was "Studies That Don't Hold Water: 30 Errors in the Environmental Impact Assessment for the Madeira River Hydroelectric Complex," released by Friends of the Earth Brazil and the International Rivers Network. The analysis critiqued the lack of coverage of impacts on biodiversity, health, flora and fauna, and populations. Many Brazilian experts and a few politicians, like the secretary of public health in Porto Velho, were quoted. No local populations had a voice. IBAMA claimed that there was insufficient information to make a decision. In response, FURNAS, the sponsoring state agency, and Odebrecht, the international company intent on building the dam, generated new studies (Switkes 2007). The IENGOs involved in contesting the Rio Madeira complex responded to the plans and benefits proposed by funders as well. For example, the Banco Nacional Desenvolvimento Economico e Social (BNDES), one of the possible partial funders of the dam that conducted viability studies, argued in 2003 that the benefits of the dams would include energy generation; the integration of transnational travel and trade through Brazil, Peru, and Bolivia, which would assist with agrodevelopment across national boundaries; and the interlinking of electrical lines across the states of Rondônia, Acre, and Mato Grosso (BNDES 2003). The International Rivers Network raised direct concerns with these developments in its report, which pointed out that the social and environmental impacts caused by them were not accounted for in the environmental impact assessment.

In the summer of 2007, the environmental license for the Rio Madeira dams was granted after several iterations of environmental assessments, dam plans, and contestation. Although the movements were able to raise concerns about the plan's impacts and, in conjunction with IBAMA, force shifts in planning, the political will of the federal administration far outweighed these voices. In addition, the legal claims made by local movements in the case of Belo Monte were not as salient as they had been in the past and were superseded by political interest in bringing plans to fruition. Although government planners have used environmental rhetoric, they have not adhered to it enough to turn to energy conservation rather than dam construction in the Amazonian rain forest. For example, a representative from Odebrecht claimed at a public hearing for the Rio Madeira that "these dams will be different from others that have been built. They will have fewer environmental impacts than other dams that have been built in Brazil." If these dams are constructed, it will indicate that the movement has lost out on one of its longest struggles and that transnational demands for energy and industrial growth have been prioritized. This recent political change reflects a decrease in accountability to citizens and an increase in importance of other governments and corporations willing to fund infrastructure development in the Amazon.

More projects are projected to be passed with the institution of IIRSA, which promises to be one of the most ambitious transnational projects ever planned in South America. Twelve South American countries agreed on IIRSA in 2000 in order to develop an infrastructure that would facilitate trade across national borders continent-wide (van Djick and den Haak 2007). The primary intention is to make regulatory and institutional frameworks for transport, energy, and telecommunications consistent across boundaries. IIRSA's rationale is based on the need to move beyond the remedying of trade boundaries and trade liberalization in order to integrate Latin America into global markets, including both North-South and South-South trade relations. Its multiple projects include roads, bridges, ports, tolls, customs facilities, and telecommunications.

Although these projects have an immense capacity to spur development that leads to South American competitiveness in international markets, if they are developed, there is little doubt that they will result in negative environmental and social impacts. The public-private partnerships (PPPs) being planned to answer the economic needs of IIRSA also incorporate mechanisms meant to resolve these potentially conflicting outcomes. PPPs take the form of long-term contractual agreements between the private and public sectors where a variety of services are bundled together and the involved parties have shared financial investment. Much of the funding will be provided by state governments or by regional banks like the Inter-American Development Bank or the Andean Investment Corporation. Other outside investors are also interested in funding projects in order to reap benefits from access to public goods and trade opportunities. Estimating the environmental outcomes of the plans is beyond the capacity of cost-benefit analyses and strategic environmental analyses (SEAs), which are the institutional mechanisms in place to account for impacts of the projects. The ADM will face greater challenges than ever confronting these projects, and even fewer institutional pathways for participation than before.

Building a New Kind of Big Science

Inklings of environmental breast cancer activism in the San Francisco Bay Area appeared in the early 1990s but did not take real shape until 1994, when the Northern California Cancer Center published an article announcing the breast cancer statistics for the San Francisco Bay Area. To the shock of the public, Marin County had the highest rates of breast cancer in the world (Clarke et al. 2002). A year later Marin Breast Cancer Watch, now called Zero Breast Cancer, was formed. It built coalitions with Breast Cancer Action, the Breast Cancer Fund, the Women's Cancer Resource Center (WCRC), and Bay View Hunter's Point Community Advocates, all of which had been founded a few years earlier.

Marin Breast Cancer Watch (MBCW) has been the primary group gaining funding for research projects in the Bay Area. In the early years of MBCW, state agencies treated the organization as radical and too left-wing to be given support. MBCW relentlessly continued its focus on environmental issues. Eventually, the organization became strongly incorporated into governmental institutions that were willing to address this topic. One mediating factor was the organizational focus on using science. The hosting of radical speakers was reduced when more organizational focus was put on involvement in science. This evolution made the group more accessible to government agencies that would be willing to fund research but not public protest. Lay participation involving the MBCW has evolved from a small-scale research project not addressing environmental causes to a large, well-funded study involving multiple EBCM organizations in the Bay Area and focused specifically on the environment. Today, MBCW runs more research collaborations than seems possible out of its small offices in a modest strip mall.

In the late 1990s, MBCW received funding from the California Breast Cancer Research Program to conduct a study of adolescent risk factors for breast cancer and housed several researchers part-time in order to do so. The Adolescent Risk Study used new methods to reconstruct life histories of survey subjects so that they might recall experiences of their adolescence (Wrensch et al. 2003). Although the study did not address environmental factors, its methodological innovation promises to help future environmental studies that have been challenged in gathering data about past events. Since that initial study, the organization has grown to engage in several other collaborations primarily centered on advancing the understanding of breast cancer risk in Marin County. For example, the Marin Breast Cancer Research Collaborative is a working group of Marin residents and research scientists including officials from the Marin County Department of Health and Human Services (DHHS), activists from MBCW, and researchers from the University of California at Berkeley, the University of California at San Francisco (UCSF), and Lawrence Berkeley National Laboratories. This group focuses on data from Marin County to develop new hypotheses about why cancer rates are higher there, test these hypotheses with data from Marin County, and provide information exchange between researchers and laypeople.

MBCW takes part in several other scientific projects. The Traditional Risk Factor and Marin Environmental Data studies use existing datasets to examine the role that factors like diet and birth of first child play in breast cancer risk. They are citizen/science alliances in the sense that research questions and methods were informed by activist concerns. In the Personal Environmental Risk Factor Study, the CDC funds the University of California at San Francisco and Lawrence Berkeley Laboratories to construct a survey tool to assess individual-level environmental exposures. The Nipple Aspirate Fluid

Collection Pilot Study was funded by the Susan Love MD Breast Cancer Foundation in 2001 to collect and analyze nipple fluid for pesticide content. This is an answer to the EBCM assertion that the concentration of toxics in the breast leads to breast cancer. However, some groups have withdrawn their support from the project. They were concerned that resultant science might communicate that women should not breast-feed, even though most findings show that the benefits are still greater than the potential drawbacks.

Working in a Larger Scope

Much as the transformation in name from MBCW to Zero Breast Cancer signifies a transition to taking a larger worldview, the EBCM in California has expanded in focus since its inception. Activism has moved from small scientific projects to the largest federally funded environmental breast cancer research project. California now contains the largest number and broadest types of lay/expert collaborations. It has also achieved the most growth in incorporating activist agendas into research policy.

In the late 1990s, EBCM activists in California encouraged the Centers for Disease Control (CDC) to conduct a study of heightened breast cancer risk factors of women in their area. The CDC focused on traditional factors like late first birth, diet, and lifestyle and found that the rates in Marin were no higher than anywhere else, that they were in fact declining, and that enough research was being done. Hearing these results, the activists encouraged CDC representatives to present them publicly so they could protest. In 1998, the CDC held a town hall meeting where activists argued that it did not matter how high the rates were, the rates were too high. Soon after, the Breast Cancer Fund requested that a special research project like that on Long Island be initiated to focus on environmental causes of breast cancer. This request motivated the NIEHS and the CDC to commission Patricia Buffler, a research professor at Berkeley, to begin organizing what would become the "International Summit on Breast Cancer and the Environment: Research Needs." This was a landmark event in 2002 that put Bay Area EBCM research on the international level. Buffler had been attending the Bay Area Cancer Study Group, where activists, scientists, and clinicians had been discussing the state of breast cancer research. They had specifically explored the role of the environment, reviewing the body of literature as a whole, in order to interpret what support existed for an environmental hypothesis. The group found that research in that area was severely lacking, if not completely misdirected. The summit was a response to these ideas and the activist concern about lack of emphasis on primary prevention.

For three days, activists, scientists, clinicians, and government officials questioned what traditional standards of proof were being used in breast cancer science. They also consulted on the present state of breast cancer research on the environment in order to determine the types of research needed in the future. Meeting attendees pointed out that current breast cancer research did not allow for action on public health issues or a precautionary approach. In the final analysis, the group recommended that "researchers and funders should consider innovative and appropriate ways to involve community members in the range of scientific studies. . . . Particular attention needs to be paid to issues of latency, area differences in incidence, migration, and measuring internal biomarkers of exposure to environmental agents" (Summit proceedings 2003).

Achievement of such a high-profile gathering was only possible because of the governmental legitimacy that activists had already achieved. At that point, activists in both Marin County and San Francisco itself had been working directly with political representatives and scientists for almost ten years. Many of the EBCM organizations in the Bay Area played influential roles in developing and conducting the conference. The two key researchers from MBCW discussed their research with the organization on one panel, while the executive director of MBCW moderated another panel focused on environmental causes of breast cancer.

The International Summit was a collaborative forum that helped shape research policy and was extended to city ordinances based on the precautionary principle. It exemplifies how the EBCM made specific changes in science to answer concerns of the affected population. The summit was guided by the idea that a community-based approach is most effective for public health because (1) "knowledge is recognized to be socially constructed and not neutral," (2) "researchers and subjects are equal partners pursuing knowledge for a common purpose," and (3) "knowledge is used to produce change and promote social equity" (Summit Report 2003: 11). The commitment to a community-based approach by the summit acknowledges the biases of science and the need to use science as a political tool.

The summit also directly addressed concerns about what activists conceived as problematic scientific norms. For example, it recommended that multidisciplinary approaches be adopted in breast cancer research on the environment. The summit's recommendation was that social scientists, toxicologists, biologists, epidemiologists, and others work together. It also recommended the development of new methods of measuring exposure. Activists had frequently pointed out that the existing breast cancer science could not claim to make substantial conclusions about lack of environmental causation because exposure measurements were so challenging.

As activists coalesced with researchers throughout the Bay Area, they began to form partnerships and constituencies that could demand both research and regulatory policy change. Research policy change came in the form of the NIEHS Breast Cancer Research Centers, and the regulatory policy came in enacting precautionary measures.

NIEHS Centers of Excellence on Breast Cancer and the Environment

In 2003, the year after the summit took place, Senator Lincoln Chafee of Rhode Island and Senator Harry Reid from Nevada introduced a congressional bill to fund centers studying environmental causes of breast cancer. Chafee was backed by a representative from the Rhode Island Breast Cancer Coalition who served on the National Breast Cancer Coalition's board and had recently engaged in work on environmental causes. Representative Nita Lowey from New York, who had lost her mother to breast cancer, introduced the bill in the House. The bill aimed to provide $150 million for five years and was meant to make breast cancer advocates pivotal in the research design and conduct. Lowey and Representative Lynn Woolsey from the Marin area intended to bring women with breast cancer and their families in to advise scientists in the conduct of the research. Despite this political push, the bill was slow to get passed. In an important demonstration of support, Ken Olden, the director of NIEHS, and officials at NCI decided to draw money from their regular budgets to fund the centers. As a result, four centers were funded in 2003, rather than the planned twelve. Each center was given $5 million a year for seven years, with a total of $150 million for all the centers.

These centers were called National Institute of Environmental Health Sciences Centers of Excellence on Breast Cancer and the Environment. They were meant to examine both biological and epidemiological aspects of breast cancer causation by studying environmental effects on molecular structures and environmental effects on puberty (NIEHS 2002). The purpose of researching these topics was to create a better understanding of how environmental exposures might cause breast cancer. Scientists, breast cancer advocates, and health-care practitioners were involved in forming the request for proposals that defines what will be studied in these centers and the methods employed. Active participation of breast cancer advocates or advisory groups was to take place in all steps of research, from planning to dissemination of information to the public. The centers attempted to reach beyond past constrictions of participatory research at the federal level by incorporating activists as principal investigators in the research and by compensating laywomen for their time. This signifies an entirely new level of appreciation and legitimation of lay perspectives.

MBCW, in conjunction with other EBCM organizations in the Bay Area, was one of the organizations that responded to the RFP for a center. Their center would involve UCSF and another research institute, Kaiser Permanente, in investigating environmental effects on mammary glands across the lifespan. Research would also look at the environmental and genetic determinants of puberty. A total of thirteen applicants applied for the four center slots. In late 2003, MBCW was one of the four groups to be awarded funding for these new centers, along with the University of Cincinnati, Fox Chase Cancer Center in Philadelphia, and Michigan State University. (See Table 7.2 for the full listing of participating Centers.) This was a major victory for activists in the area. Along with MBCW, women from the Bayview Hunter's Point Project Community Advocates, the Breast Cancer Fund, and other EBCM organizations are involved in working with researchers to study these topics.

The RFP stated that the primary goal was to establish a "network of research centers in which multidisciplinary teams of scientists, clinicians, and breast cancer advocates work collaboratively on a unique set of scientific questions that focus on how chemical, physical, biological, and social factors in the environment work together with genetic factors to cause breast cancer." This lay/expert collaboration is the most formalized, institutionalized, and governmentally supported of all EBCM projects. However, whether activists will feel sufficiently included is still a question. During the first stages of the project, many laypeople felt that their opinions were not actually being considered, and contention between scientists and laywomen arose during annual conferences (McCormick and Baralt 2008). More recently, some activists on the advisory committee for the national coordinating committee have felt that their involvement is well received and makes a difference. These projects, and the following regulatory policy changes, indicate that the EBCM has created

TABLE 7.2 BCERC COLLABORATORS AND COMMUNITY PARTNERS

	UCSF	University of Cincinnati	Fox Chase Cancer Center	Michigan State University
Expert collaborators	California Department of Health Services, Kaiser Permanente of Northern California, Lawrence Berkeley National Laboratory, Marin County Department of Health and Human Services, Roswell Park Cancer Center, University of Michigan	Cincinnati Children's Hospital Medical Center	Mt. Sinai School of Medicine, the University of Alabama	N/A
Community partners	Zero Breast Cancer, Bay Area Breast Cancer SPORE Advocacy Group, Bayview Hunters Point Health and Environmental Assessment Task Force, Breast Cancer Fund, Community Health Agency	Breast Cancer Alliance of Greater Cincinnati, Patterns, Inc., Pink Ribbon Girls, Susan G. Komen Breast Cancer Foundation, Greater Cincinnati Affiliate, American Cancer Society, Cincinnati Area Office, Breast and Cervical Cancer Screening Project, Greater Cincinnati Occupational Health Center, Lower Price Hill Women's Wellness Group, National Breast Cancer Coalition, the Wellness Community, YWCA Breast and Cervical Health Network	The Renaissance University for Community Education of the Harlem Children's Zone Project, Girls, Inc., New York City Parks Foundation, Community Science Specialists, Share	Faith Access to Community Economic Development, Susan G. Komen Breast Cancer Foundation, Michigan Environmental Council, American Cancer Society, Great Lakes Division

an avenue through which it can influence scientists and policy makers more than in the past.

Passing a Precautionary Principle Ordinance

As activists in California worked on gaining support for new research legislation, others gained political connections that led to the passage of local regulatory policy. One of the first such attempts at regulatory change originated with Breast Cancer Action becoming a part of the Pollution Prevention Advisory Committee initiated by the Department of Toxic Substances Control in the California Environmental Protection Agency. The committee was formed by the Sierra Club and was meant to facilitate the adoption of voluntary pollution-prevention measures by industry. Although the composition of the committee included environmental, public health, government, and industry representatives, the BCA member was one of only two grassroots activists. The P2 Advisory Committee first chose to address the oil industry, which is the largest producer of hazardous waste in the state and has many facilities located in underprivileged communities. In order to facilitate community/industry partnerships that would lead to pollution prevention, the committee pointed out the need for industry to share information about its waste creation and disposal. However, when security measures against terrorist threats were instigated by the September 11, 2001, attacks, the project was closed down.

After this initial failure, activists achieved their first important regulatory policy change with the Berkeley healthy building ordinance, which would eliminate toxic polyvinyl chloride and formaldehyde. This ordinance would also work to increase public awareness of the carcinogenicity of industrial pollution by forcing the city to distribute related fact sheets to hospitals and other health sites. Government representatives would act as a part of the Association of Bay Area Governments task force on dioxin prevention. The city also agreed to ban pesticides in public places and promote internal purchasing of nonbleached paper. Soon after, activists in the Bay Area were able to pass a major city ordinance in San Francisco and put it on the agendas of other city councils, like Berkeley and Oakland (Breast Cancer Fund 2003a). It stipulated that all government agencies had to purchase environmentally friendly or noncarcinogenic products when they were available:

> Utilizing the Precautionary Principle to select products and services that minimize negative impacts to human health and

> the environment will use San Francisco's significant purchasing power to create markets for alternative products, thereby driving manufacturers to engage in research and development towards the production of additional innovative alternative products. (Breast Cancer Fund 2003b)

This achievement was pushed forward by a group of organizations in the Bay Area called the Precautionary Principle Working Group that had been meeting with government officials for more than a year. The majority of these groups were EBCM organizations, and others were only environmentally focused. The ordinance was passed on June 24, 2003, following implementation of the precautionary principle (PP) in the European Union policy (Commission of the European Communities [CEC] 2000). Activism on the PP had already been taking place for several years in the EU (Gremmen and van de Belt 2000). EBCM activists in the Bay Area had been leveraging these EU achievements in their claims that better regulatory policy should be implemented in the United States as well.

Meanwhile, other EBCM activists in California pushed legislation at the county level. For example, Marin Breast Cancer Watch had a close connection with Pesticide Watch in their area. Pesticide Watch conducted a campaign with the National Association against the Misuse of Pesticides that analyzed the use of pesticides in Marin County for MBCW. Finding that there were incredibly high levels of use for a county with no agriculture, MBCW formed the Marin Beyond Pesticides Coalition, a new organization that focused on reducing community-level use. It was ultimately successful, as one activist recounted:

> It took a year and a half. In December of 98 we got the county to pass the IPM [Integrated Pest Management] ordinance and they reduced the use of pesticides by 75 percent by the target years which is . . . three years from now. And [they] set up a commission which is now meeting [and they] have coordinators and to notify people when they spray. And notify them where they are spraying because nobody knew before. It was part of the whole secretive chemical industry. As long as people don't know, they're not going to protest.

As this story portrays, MBCW was successful in changing local regulation and use of toxics and also in increasing public awareness about their use. These changes signify an increased sensitivity to toxic exposures and the beginnings of shifts in patterns of consumption.

Incorporating a precautionary approach into governmental practice is possibly the ultimate goal of the EBCM. A precautionary approach minimizes exposure to cancer-producing chemicals and therefore reduces breast cancer risk. Consequently, policy changes catering to this demand signify the incorporation of the EBCM discourse into governmental practice.

Growth, Change, and Impact of Democratizing Science Movements

DSMs respond to the scientific, economic, and political climate in which they work. There does not appear to be a particular trajectory that these movements follow. Therefore, in order to understand how future movements may play out, researchers must assess how engagement in research works in conjunction with the broader social landscape and the other forms of activism used by the movement. For example, in the case of Belo Monte, when the international anti-dam movement became less involved in contestation, the movement relied more heavily on collaboration with scientists to stop the dam. When the EBCM was able to gain access to a major sympathetic political representative who would push the NIEHS centers forward, activists focused on this opening rather than other activities. Developing roles of research activities in movements is therefore intimately interconnected with other movement activities and political opportunities. What these movements are able to achieve reflects, in part, the intention and resources of each movement, but possibly more important, it indicates the shifting landscape of politics and science. Movement outcomes indicate the agenda of government actors and their accountability to local concerns. Therefore, they can reflect how governments and movements can create more synergy on controversial issues.

In particular, differences between these movements indicate the importance of national context, the extent to which science is a tool for large corporate or government interests, and the participatory conditions of scientific institutions. These institutions can either create democracy or exacerbate disenfranchisement. As a result, the EBCM and the ADM have achieved different levels of research and regulatory policy change. The EBCM has been able to gain millions of dollars in funds for environmental breast cancer research and some regulatory change. On the other hand, the anti-dam movement has made major regulatory change in specific cases of dam construction and has possibly contributed to the overall energy model.

The recent developments of the ADM and the EBCM represent long-term struggles that have yet to be resolved. For the EBCM, the duration of the NIEHS centers project and the degree to which they are participatory reflect the success of the movement in demanding advanced participation. In the past, such research has been constrained by the short period of time that researchers and activists have had to work together. CBPR is notoriously difficult and lengthy, so the seven-year time period of the centers allows time for the projects to grow to resolution. In addition, the degree to which advocates have been involved in the centers appears to be an advance over past strategies.

The ADM has been remarkably successful in defeating what would be the largest dam in Latin America and, possibly more important, in the Amazon. Other local movements have not been as successful, and it is clear that the involvement of transnational social movement organizations and high-profile researchers has boosted the ADM's success. Transnational norms about protection of the Amazon are possibly also at play in this case. The recent passage of the Rio Madeira dams that were proposed as the alternative to Belo Monte suggests that the weaker activism in that area was not able to challenge governmental plans in the way that these historical actions were.

Both the EBCM and the ADM are now engaged in large-scale, long-term struggles over some of the most contentious and charged issues they have faced yet. As globalization of large economic interests continues and as biotechnologies gain in profitability, new institutions of democracy must be developed to mediate controversies over these developments. Clearly, global discourses, such as the precautionary principle and sustainability, have been absorbed by these movements and translated into government. However, institutional change is more difficult. These cases make it clear that sufficiently participatory institutions do not yet exist. However, these movements have begun to provide models of what such institutions must look like in order to be effective. On the basis of their ability to engender this much change in some of the most deeply entrenched ideas of development and health and to counter some of the largest economic interests in the world, these movements portend radical change in the institutions of science in the future and yet potentially also the continued co-optation of social movements.

8 A Case for Making Science Accountable

Democratizing science movements are emerging around the world. Genetically modified organisms, nanotechnology, cloning, and many other ethically charged scientific advances are surrounded by conflict between their producers and consumers. These debates beg important questions about ownership, government oversight, and the role of civil society in decision making about scientific advances. Reflecting the diverse topics they contest, activists work under multiple conditions and are consequently able to achieve various goals. As a result, these movements are sometimes successful, but frequently they fail. This book has made a broad, sweeping attempt to review what these movements are and their outcomes. On the basis of this analysis of DSMs—what they indicate about relationships between science and society, the role that corporate control plays in those relationships, and what types of participatory projects are necessary to overcome imbalances in power in these relationships—I begin to answer questions regarding how these movements function, when they succeed, and what makes them fail.

DSMs draw attention to the broad social context in which science functions. These activists clearly demonstrate how scientization shapes their lives. They point to the intimate connection between science and politics, and how opaque scientific language and insulated practices intimately affect nonexperts. Activists draw attention to the role that corporate interests play in scientization, or how such interests use research findings and scientific paradigms as a way to subtly influence politics and consumption. These movements connect the importance and utility of participatory structures to possibilities for deepened democracy by demonstrating that research and scientific development are important in political institutional arenas. Often, however, their work is hindered by superficial participatory mechanisms that give the appearance of deliberation but result in few outcomes for which movements are fighting.

Social Context of Science

As of the early twenty-first century, most of us take for granted that science is central to politics. In fact, we become outraged when policy is based on norms or ideology instead, or when we think that policies have not adequately accounted for scientific findings. In many ways, we, at least subconsciously, view science as a protector of democracy, equality, and fairness. For example, many countries are disgusted by American disregard for the vast majority of science that supports a link between industrial production and global warming. While they expect or hope for a set of just policies based on science, America continues to use a normative structure that values economic growth over protection of health and the environment. Another example is recent controversies over intelligent design. Even most conservatives were opposed to basing classroom education regarding the history of mankind's development on a religious or normative approach rather than a scientific one that uses evolution.

Although we believe that science is central to politics, we often simultaneously disregard the influence that politics—in the form of identity, money, or power—has on the development of science. Before a scientific study even enters the political debate, it is shaped by the identity of the individuals doing the research, the financial support that makes it possible, and the power structures that have created methods, areas of research, and topics of study in that field. These are the subtle or sometimes more obvious ways in which Schattschnneider's conception of scope gets manifested in movement contests over science. The privatization of science limits the scope of contest, but the development of a boundary movement or democratizing science movement can surpass this limitation by bringing new forms of power and knowledge into social movement struggles. Democratizing science movements have emerged to alert us to problems with the way that science is used in politics, as well as the way it is constructed.

Along with many other movements, the environmental breast cancer movement and the anti-dam movement challenge the science that has already been incorporated into political decisions while simultaneously constructing knowledge that takes new issues, identities, and concerns into account. Because of this dual agenda, these movements must use a number of tactics. Democratizing science movements mobilize resources and take advantage of political opportunities, even though these tactics are not always obviously tied in with science. They relate directly to challenging the state by making political connections, developing new institutional forms, and reshaping governmental norms. These movements have also developed new tactics that connect science and politics. The lay/expert collaboration is the most central of these tactics. There are four main components of the lay/expert collaboration: being educated in technical knowledge by researchers, acquiring new forms of legitimacy and authority through engaging with scientists who serve as a new type of activist, developing new research and critiques of existing research with scientists, and presenting new science in official settings like workshops and governmental meetings. These collaborations result in DSMs being able to translate research findings to the public in a way that further stimulates public engagement in science and technology.

Participatory projects, like the ones discussed in this book, are burgeoning all over the world. In many cases they are driven by an active civil society and the growth of a social movement. Many researchers choose to focus solely on the participatory research project or the newly developed governmental institution rather than the broader social landscape from which they emerge. Although this type of analysis provides one side of the story, I argue that in order to understand how democracy is deepened and science is made more accountable, we must also examine the groups that stimulate the evolution or, often, devolution of participatory practices. By looking at them, we can better understand why and how they do this, and what they can really achieve.

Democratizing science movements work in the realms of the scientific, political, and public because of particular characteristics of these three realms. First, these movements argue that science and technical knowledge are biased and often not reflective of their experience. Because of that, they attempt to generate technical codifications that include their knowledge. Second, despite its lack of objectivity, expert opinion is institutionalized in governmental decision making and consequently marginalizes lay perspectives and regular citizens. Often the funding for studies comes from corporate sources that generate biased data and push governmental actors to keep research agendas focused on certain trajectories. Finally, since much of the public views science as unquestionable, these movements must also change public opinion by drawing attention to less prominent scientific arguments, gaps in research, and political interests that shape what we know. Although democratizing science movements have made many advances, they are also limited by facets of science itself, such as the length of time necessary for research or trajectories of research that are difficult to change, and by powerful interests that choose to disregard expertise with which they do not agree.

Democratizing science movements contest and control expert knowledge to reempower the lay citizen in the face of scientization. One of the most important stimulants to the rise of democratizing science movements is the role of private interests and corporate control over expert knowledge. A growing amount of power has been concentrated in the hands of experts and their corporate funders. Democratic practice has lagged in managing power differentials between these experts and citizens. Activists' intent is not to stop the development of science and technology but rather to reembed it in good public opinion. Despite variations in democratic development in Brazil and the United States, governmental use of expert knowledge supports industrial interests. For example, the private interests that intend to build a dam or begin selling a new chemical conduct their own research. While some view this as a positive practice of placing the burden on industry, it also gives industrial sources much more power than the public. The government agencies that use these reports and give little access to the public to comment on them only exacerbate the insulation of expert knowledge generated by private sources.

In the case of dam building, companies hire experts to assess environmental and social impacts. These reports are evaluated by yet another body of experts at the state environmental agency who then hand their report to another technocrat who makes the final decision. Three groups of experts examine the pros and cons of the dam without necessarily obtaining the input of a local person who might be affected. Although the first environmental impact assessment may be commented upon by the local community, community members often do not have that opportunity. When they are able to participate in a public hearing, their input is not formally considered. When the anti-dam movement began, there was very little space for the participation of affected people. Since then, it has achieved significant forms of institutional participation, such as advising political representatives, the construction of environmental impact assessments, and delineating resettlement packages. Misinformation or lack of information reduces local populations' ability to assure their rights in the environmental hearing process and further allows for the prioritization of industrial interests over social and environmental concerns of affected communities. Lay/expert collaborations countered these problems by sidestepping officials who gained financially from dam-building companies and by making negotiation a group process. Activists often also pointed out that this alliance was problematic for both the economic growth of the country and the environmental and social impacts of development.

Scientific institutions where government and corporate institutions share research funding and paradigms are boundary organizations where similar interests, trajectories, and paradigms are advanced. Although the concept of boundary organizations and objects helps theoretically demonstrate how government and corporate institutions intersect, it is methodologically challenging to find data that specifically link the two. In Brazil, officials and activists spoke of the ways in which officials gained financially from dam construction. Similarly, in the United States, lobbying by large pharmaceutical companies and chemical manufacturers results in money for government representatives. In addition, pharmaceutical companies gain from government-funded scientific advances, while private dam funders profit from the provision of public resources, like water and land.

Two Sides of a Barbed Fence: Challenges for Experts and Laypeople

These two movements face very different limitations and use varying tactics to overcome them. Engaging in science also has direct effects on movements. First, there are significant internal scientific barriers that democratizing science movements must confront and are not easy to change. These challenges can impede collaborations and make the outcomes of research less than activists had hoped for. Second, movements must sacrifice valuable time and resources in order to engage in scientific projects. As has been shown in past research on community-based participatory research (Minkler 2000), conducting research is a time-consuming process, especially for those with little training. Finally, in many cases, even when movements are able to change science, this achievement is undone or goes unrecognized by larger political and corporate interests. When this happens, the time and energy put into collaborations with experts may seem like a waste. Internal scientific barriers include norms that circumscribe or motivate study. One researcher explained:

> People are hoping to have a big hit so scientists are developing their own companies to try and enter into partnerships with industry. I think that activists being more involved in that process brings everything back down to scale. Where you know the people who suffer from the disease really are present day to day in people's faces. That makes the atmosphere less rarified and less intellectualized.

But "hitting it big" seemed to be more difficult for researchers studying environmental causes of breast cancer than for those taking a biomedical approach. Consequently, a part of making science more democratic was changing the priorities of science and shifting the focus on serving industrial interests to serving affected communities. However, taking an environmental approach made researchers more marginalized in the breast cancer science community and more challenged by more traditional researchers. One researcher said:

> It is hard to get peoples' attention when you are from _____ and doing research about environmental causes of breast cancer. It's hard to break through and get people's attention, scientists' attention. You go through a kind of hazing where you have to establish talking about what you're doing, your credibility with each new member. You don't have on your nametag Harvard University. You really have to go through a process of demonstrating your credibility.

Researchers engaging in lay/expert collaborations with anti-dam activists faced the same normative barriers to engaging with nonexperts that researchers in the United States dealt with. One researcher explained:

> In this Institute many professors have this work with social movements. It is not a unique characteristic because there are other universities like this. It is called extension. The Brazilian constitution says that learning and research and extension are supposed to be connected and go together. So it is supposed to be legitimate. In many cases it is professors

> whose work is not very good that are doing this kind of work. When you are not a very good scholar you cannot do this kind of work.

Researchers often lose credibility when they engage in lay/expert collaborations. This was true in many of the previously discussed collaborations. Advances in science generated by lay/expert collaborations were relatively basic but were the first steps in creating more just political outcomes for dam-affected people.

On the other side of collaborations, as activists became well versed in scientific vocabulary and methods and became more accustomed to working with researchers, they realized that scientific advances face many constrictions and much fallibility. One EBCM activist who had served on numerous scientific review boards said:

> It's too frustrating for me to read these reviews because for every study there's another study that says it's [an environmental causation hypothesis] not true. On top of that there's another study that says it's true. On top of that there's another three studies that say it isn't true. . . . So to me it's like science has to be there but it's real slow and I don't know if they're always comparing apples to apples. I'm finding that the science part of it, I can't wait for science any more.

Another activist said:

> The problem is that it's [science] so time intensive. The research particularly takes a lot of staff time and energy to do all that stuff. . . . We've tried to ride the wave of balance around those things—research or advocacy. We're really realizing now that we need to do a lot more education around what we learned through the research, translation of everything back to the community, educating about the process of doing research and why it's so complex and why it takes time and why there are no simple answers.

This realization about the slowness of science and constrictions of scientific certainty led many activists to focus on the need for political action along with scientific work. One ADM activist who had engaged in both politically successful and defeated collaborations said, "We work with the professors, but sometimes developing new research does not work and we have to protest. The environmental impact assessments are not always enough." Challenges within science and to movements make it difficult to achieve the change that social movements seek. These challenges raise some new issues regarding theories of co-optation. Co-optation has traditionally been clearly defined as a process in which more powerful interests assert their claims over less powerful social actors as they work together under the guise of achieving change. In these cases, scientists and activists, the two main groups of social actors in DSMs, are limited by institutional structure and the norms of science. Co-optation can be far from purposeful; rather, it can be a default outcome that those involved did not anticipate or hope for.

Overcoming Limitations

Democratizing science movements use some unique tactics to overcome obstacles particular to engaging in science, as well as other methods typical in movements more generally that can resolve some of their limitations. The goals of DSMs are diverse, but some of the most common are reframing science, changing scientific paradigms and processes, and shaping planning and regulatory policy. The cases in this book have shown that like these goals, the tactics necessary to achieve them are interdependent and interrelated. Therefore, this book confirms the importance of political opportunities and advances that theoretical approach by pointing to the role of science within them. More specifically, it shows that political dependence on science is a tool for disenfranchisement and resultant social movement formation. Therefore, political pathways for participation in research are a new form of democratic participation that movements demand. In this way, DSMs are shaped by the form of the state, as well as the economic actors to which it is connected. This research also responds to scholars who have recently raised the question of the importance of the state versus the power of transnational economic interests. These cases show that research is a mechanism through which corporate interests legitimate themselves in the eyes of the state but weaken government institutions by undermining their credibility. These moments are also possibilities for co-optation of movements. As movements arise to contest corporate influence on state regulatory and research policies, they can either be accommodated or marginalized.

Movements therefore need access to politicians who can choose to specifically advance movement demands. Often this means introducing a new discourse or set of exposures about which policy makers should be concerned. Movements may take this direct approach without going through scientific channels first. Even in DSMs, they continue to depend on connections with political representatives. Sometimes they gain these connections through their work with experts. At other moments, political connections emerge from the movements themselves. In either case, these connections must supplement lay/expert engagement in order to advance policy measures specifically. Activists may be able to advance research and scientific paradigms without these connections.

The success of the EBCM, the ADM, and other DSMs is surprising if one considers the vast possibilities for co-optation or failure. They have been able to achieve broad-ranging outcomes that affect both science and society. Keys to these struggles are sympathetic experts, open political institutions, and mass mobilization. Science runs throughout these venues, and so these movements challenge it as they challenge the institutions and practices that use it.

DSMs are able to achieve major outcomes through contesting and shaping science. The persistence of such methods attests to their importance. In an age of increasing public distrust of government and science, public involvement has been a tool that teaches laypeople about science and sensitizes researchers to the concerns of the public. Collaborations introduced by social movements help overcome the specialization of roles inherent in a bureaucratized society. Research about the environmental causes of breast cancer and the impacts of dams is important not only for its effects in the particular areas of science that it investigates but also for its role in changing attitudes and opinions of involved activists and scientists.

These collaborations and the movements behind them signify a new form of democratic practice fundamental to a technological society. In order to understand how democracy can be maintained or advanced as science and expert knowledge become increasingly important in many countries, we must take note of these informal and formalized partnerships between laypeople and experts. The lay/expert collaboration is a place where new conceptions of breast cancer causation, the impacts of dams, and the functioning of democracy were developed and led to shifts in science and policy. In a climate of scientific debate, governmental uncertainty, and methodological inadequacy, activists began to insert their perspectives. In each locale, unique collaborations have been created with experts to alter scientific procedure. Despite their differences, one outcome has been strikingly similar. Scientists have lost their skepticism. Activists have dropped their reticence. Pathways have been built across the lay/expert divide, and science has marched in new directions. In this sense, democratizing science movements can be a part of making Kuhnian paradigm shifts (Kuhn 1962). By bringing to light the shortcomings of science, they move it in new directions, most of which were originally unimagined or at least unexplored.

Although the citizen/science alliance generated the most immediate outcomes of democratizing knowledge, the researcher educator, the researcher activist, and the collaborative forum were all critical. When researchers played the roles of educator and activist, they offered legitimacy and information that allowed activists to effectively engage in CSAs and ultimately achieve the democratization of knowledge. The researcher activist was the most common of these forms, and the collaborative forum was the most important. The researcher educator was oddly absent in the EBCM despite the strong need for activists to be oriented to the scientific methods and issues they faced in lay/expert collaborations. All these formats led to important changes in scientific values and practices, as well as policy.

The typology of collaboration offers a theoretical toolbox that can be applied to other social movement case studies to elucidate how knowledge is transferred and how democratizing science movements work. Each one of these forms served different purposes and was used to varying degrees by each movement. These differences in themselves reflect inequalities in access to information and legitimacy. They also demonstrate an important point about participatory projects—that ramifications of interactions between laypeople and experts extend far beyond the confines of scientific development or political structures. The interactions outside such institutions are as critical as they are within them.

Seeing beyond These Cases

Because science and science-based policy making have grown in universality, it is likely that democratizing science movements will as well. Resolving conflicts between traditional or local knowledge systems and their Western counterparts is more than simply a matter of how we understand our reality; rather, it is about whose reality is legitimized. It is also ultimately about hypothesizing the prevention of possible environmental ruin and degradation of democracy. A hypothesis about the prevention of this degradation must necessarily intersect with many different intellectual discussions, such as development, the social construction of knowledge, and not least the power of the "countermovement" (Polanyi 1977) meant to protect the disempowered from industrial capitalism.

Many social movements attempt to counter the imposition of Western science and knowledge systems. These activists cannot be classified by topic of contestation but rather by the strategies they adopt and what they attempt to achieve. There are an endless number and types of corporate or government practices to be contested, and of institutions and processes that induce contestation of them. This will especially be the case as transnational corporations implement similar strategies to insert their interests into governmental systems that attempt to slow degradation of their lands and peoples. Western paradigms of expertise deny knowledge outside the biomedical model and create historical amnesia in their place (Rosenberg 2000). Therefore, legitimizing nonexpert perspectives is a counterforce to increased stratification linked to industrialization.

Because they are based on science, knowledge, and culture, contestation and democratization of knowledge are a new social movement tactic used to achieve social justice. By extension, we might wonder if the concepts of scientization or democratizing science movement can be applied to other contexts around the world. Cases with which these concepts resonate are arising in a number of countries and raising questions about the role of movement reframing, the identity of movement actors, and the controversiality of new technologies. In a manner similar to the DSM cases in this book, in Northern Europe, scientists, activists, and planners have begun to use the scientific framework of the "ecological corridor" as a way to come together around land-use planning in a scientifically sound and socially justifiable way (Van der Windt and Swart 2008). Likewise, a number of different organizations in the United Kingdom are forming new movements around issues of human rights and genomics by bringing together science and social issues (Welsh, Plows, and Evans 2007). In other instances, however, allegiance to science can also hinder potential activists from engaging in movement struggles or engaging with movement frameworks, as is the case with women scientists who shirk work with feminist activists (Phipps 2006).

The cases in this research offer new ideas for building successful environmental movements in other contexts. Environmental movements are the classic case to contest expert knowledge, since related policy is clearly based on research. Many other environmental movements around the world are using the same tactics as the environmental breast cancer and anti-dam movements. Understanding these innovative methods may contribute to the development of more effective environmental movements in many settings.

Democratizing science movements are not just about science; they are also about technology, power, and politics. With the growth of science and technology, these three components of life have meshed with one another to the point that they are almost indistinguishable. For example, genetic testing is simultaneously about the political economy of shaping populations, reinforcement of preexisting stratification, and simply increasing health outcomes. Life support systems and medical interventions that extend life make the importance of science and technology all the more evident. The simple existence of these medical technologies has wrested decision making about life and death away from the individual and put it in the hands of experts. At the same time, politicians battle one another for control over the use of these interventions. But DSMs attempt to reclaim the power of the citizen by taking decision-making power out of the hands of scientists, doctors, lawyers, and other experts, who are at least partially controlled by the state and corporations. As technology grows in shaping and controlling our lives, these movements will become more and more common. Their growth makes it clear that there should be a newly institutionalized role within scientific development—the citizen role. In order to resolve the growing tension between civil society and science and create increased "synergy" (P. Evans 1997) between citizens and the state, political institutions must guarantee that those most affected by science, technology, and their related political decisions have a place at the table.

Appendix: Abbreviations

Brazil

ANA	Agencia Nacional de Aguas, National Water Agency
atingido	Portuguese word for dam-affected person used by the movement and government
CEMIG	Companhia Energetica de Minas Gerais, Energy Company of Minas Gerais
CRAB	Comissão Regional de Atingidos por Barragens, Regional Commission of People Affected by Dams
EIA	Environmental impact assessment
FIAN	FoodFirst Information and Action Network
GT Energia	Grupo de Trabalha de Energia, GT Energia, Energy Working Group
IBAMA	Instituto Brasileiro do Meio Ambiente e dos Recursos Naturais Renováveis, National Institute of the Environment
IPPUR	Instituto por Planejemento Publico Urbano e Regional, Institute of Public Urban and Regional Planning
IRN	International Rivers Network
MAB	Movimento dos Atingidos por Barragens

MDTX	Movimento Pelo Desenvolvimento da Transamazonica e Xingu, Movement for Development of the Trans-Amazon and Xingu River
MST	Movimento Sem Terra, Movement of Landless People
UFMG	Federal University of Minas Gerais

United States

ACS	American Cancer Society
BCA	Breast Cancer Action
BCF	Breast Cancer Fund
CDC	Centers for Disease Control
CDMRP	Department of Defense Congressionally Directed Medical Research Programs
FDA	Food and Drug Administration
EDH	Endocrine disruptor hypothesis
EPA	Environmental Protection Agency
EWG	Environmental Working Group
HBAC	Huntington Breast Cancer Action Coalition
LIBCSP	Long Island Breast Cancer Study Project
MBCC	Massachusetts Breast Cancer Coalition
MBCW	Marin Breast Cancer Watch
NCI	National Cancer Institute
NIEHS	National Institute of Environmental Health Sciences
NIH	National Institutes of Health
PAC	Public Advisory Committee (for Silent Spring Institute)
PAH	Polycyclic aromatic hydrocarbon

References

Ackerman, John A. 2005. *Social Accountability in the Public Sector: A Conceptual Discussion.* Washington, DC: The World Bank.

Agencia Nacional de Aguas. 2003. "Informações Hidrológicas." (http://www.ana.gov.br/gestaoRecHidricos/InfoHidrologicas/default2.asp).

Agrawal, Arun and Krishna Gupta. 2005. Decentralization and Participation: The Governance of Common Pool Resources in Nepal's Terai. *World Development* 33(7): 1101–1114.

Alario, Margarita. 1994. Environmental Destruction and the Public Sphere: On Habermas's Discursive Model and Political Ecology. *Social Theory and Practice* 20:327–335.

Alcoa. 2004. *Alcoa Is Focused on Customers: 2004 Annual Report.* Pittsburgh, PA: Alcoa.

Allen, Barbara. 2003. *Uneasy Alchemy: Citizens and Experts in Louisiana's Chemical Corridor Disputes.* Cambridge, MA: MIT Press.

Altman, Roberta. 1996. *Waking Up/Fighting Back: The Politics of Breast Cancer.* Boston: Little, Brown and Company.

Alvares, Claude. 1992. Science. In *The Development Dictionary,* Wolfgang Sachs, editor, London: Zed Books.

American Cancer Society. 2007. *Cancer Facts and Figures, 2007.* Washington, DC: American Cancer Society.

Angell, Marcia. 2004. *The Truth about the Drug Companies: How They Deceive Us and What to Do about It.* New York: Random House.

Anglin, Mary K. 1997. Working from the Inside Out: Implications of Breast Cancer Activism for Biomedical Policies and Practices. *Social Science and Medicine* 44:1403–1415.

Araujo, Maria L. 1990. *Na margem do lago: Um estudo sobre sindicalismo rural.* Recife, Brazil: Fundação Joaquim Nabuco.

Arora, Vikas. 1997. FDA Policymaking in Light of Scientific Uncertainty: The Bittersweet Approval of Aspartame. Food and Drug Law Website. January 24, 1997. (http://leda.law.harvard.edu/leda/data/185/varora.html).

Aschengrau, Ann, David Ozonoff, Patricia Coogan, Richard Vezina, Timothy Heeren, and Yuqing Zhang. 1996. Cancer Risk and Residential Proximity to Cranberry Bog Cultivation in Massachusetts. *American Journal of Public Health* 86:1289–1296.

Astrazeneca. 2004. *Thinking Performance: Annual Report and Form 20-F.* London: Astrazeneca.

Austen, Ian. 2008, April 16. Canada Likely to Label Plastic Ingredient. *New York Times.*

Backstrand, Karin. 2002. *Civic Science for Sustainability.* Berlin, Germany: Paper presentation for the Conference on Human Dimensions of Global Environmental Change.

Baiocchi, Gianpaolo. 2001. Participation, Activism, and Politics: The Porto Alegre Experiment and Deliberative Democratic Theory. *Politics and Society* 29(1): 43–72.

Barrow, Craig S. and James W. Conrad. 2006. Assessing the Reliability and Credibility of Industry Science and Scientists. *Environmental Health Perspectives* 114(2): 153–155.

Batt, Sharon and Lia Gross. 1999. Cancer, Inc. *Sierra Magazine* September/October 1999. (www.sierraclub.org/sierra/199909/cancer.asp).

Beck, Ulrich. 1992. *Risk Society.* London: Sage.

———. 1995. *Ecological Enlightenment: Essays on the Politics of the Risk Society.* Atlantic Highlands, N.J.: Humanities Press.

Bekelman, Justin E., Yan Li, and Cary P. Gross. 2003. Scope and Impact of Financial

Conflicts of Interest in Biomedical Research: A Systematic Review. *Journal of the American Medical Association* 289:454–465.

Bell, Daniel. 1999. *The Coming of Post-industrial Society.* New York: Basic Books.Benford, Robert D. and David Snow. 2000. Framing Processes and Social Movements: A Review and Assessment. *Annual Review of Sociology* 26:611–639.

Berkamp, Ger, Matthew McCartney, Pat Dugan, Jeff McNeely, and Mike Acreman. 2000. *Dams, Ecosystem Functions and Environmental Restoration.* Thematic Review II.1 prepared as an input to the World Commission on Dams. Cape Town, South Africa: World Commission on Dams. (www.dams.org).

Bermann, Celio. 2002. *Energia no Brasil: Para quê? Para quem?* Rio de Janeiro, Brazil: Instituto Brasileiro de Análises Sociais.

Bermann, Celio and Osvaldo S. Martins. 2000. *Sustainable Energy in Brazil.* Rio de Janeiro, Brazil: FASE. Instituto Brasileiro de Análises Sociais.

Bernal, J.D. 1939. *The Social Function of Science*. London: Routledge.

Biagioli, Mario. 1999a. Introduction. In *The Science Studies Reader,* Mario Biagioli, ed. New York: Routledge.

———. 1999b. Aporias of Scientific Authorship: Credit and Responsibility in Contemporary Biomedicine. In *The Science Studies Reader,* Mario Biagioli, ed. New York: Routledge.

Bimber, Bruce and David H. Guston. 1995. Politics by the Same Means: Government and Science in the United States. In *Handbook of Science and Technology Studies,* Sheila Jasanoff, ed. New York: Sage Publications.

Bohman, James and William Rehg, eds. 1997. *Deliberative Democracy: Essays on Reason and Politics*. Cambridge, MA: MIT Press.

Böhme, Gernot and Nico Stehr, eds. 1986. *The Knowledge Society: The Growing Impact of Scientific Knowledge on Social Relations*. Dordrecht, Holland: D. Reidel Publishing Company.

Bourdieu, Pierre. 1992. *The Logic of Practice*. Palo Alto, CA: Stanford University Press.

Braga, Flavia. 2000. *Brazil's Movement of Dam-affected People and the World Commission on Dams*. Background pPaper for the WCD Assessment. World Commission on Dams. (http://www.dams.org/contact.htm).

Braun, Lundy. 2003. Engaging the Experts: Popular Science Education and Breast Cancer Activism. *Critical Public Health* 13(3): 191–206.

Brazilian Development Bank. 2003. *Premeiro Seminario Internacional de Cofinanciamento*. Rio de Janeiro, Brazil: Brazilian Development Bank (BNDES/CAF).

Breast Cancer Action Website. 1993. Position Paper on Breast Cancer Advocate Involvement in the Research Process, Newsletter #19. San Francisco, CA: Breast Cancer Action. (http://bcaction.org/index.php?page=newsletter-19d).

———. 1999. Breast Cancer Action Counters Direct-to-Consumer Ads for Tamoxifen, Press release. San Francisco, CA: Breast Cancer Action. (http://bcaction.org/index.php?page=991025-2).

Breast Cancer Fund. 2003a. Bay Area Working Group on the Precautionary Principle, Press advisory. San Francisco, CA: Breast Cancer Fund.

———. 2003b. Precautionary Principle Resolution. San Francisco, CA: Breast Cancer Fund. (http://www.breastcancerfund.org/pp_resolution.htm).

———. 2007. The Healthy Californians Biomonitoring Program SB 1379. San Francisco, CA: Breast Cancer Research Program. (http://cdmrp.army.mil/bcrp).

Brenner, Barbara. 2000. *The Breast Cancer Epidemic*. Paper presented at the American Public Health Association Conference. Boston, MA: American Public Health Association.

Bridges, Andrew. 2007, June 30. FDA officials criticized for secrecy. *USA Today*.

Brody, Julia G. 2004. Hair Dyes and Cancer. *New York Times* op-ed.

Brody, Julia G., Ann Aschengrau, Wendy McKelvey, Ruthann A. Rudel, Christopher Schwartz, and Theresa Kennedy. 2004. Breast Cancer Risk and Historical Exposure to Pesticides from Wide-Area Application Assessed with GIS. *Environmental Health Perspectives* 112(8): 889–897.

Brody, Julia G. and Ruthann A. Rudel. 2003. Environmental Pollutants and Breast Cancer. *Environmental Health Perspectives* 111(8): 1007–1019.

Brody, Julia G., D. J. Vorhees, S. J. Melly, S. R. Swedis, P. J. Drivas, and Ruthann A. Rudel. 2002. Using GIS and Historical Records to Reconstruct Residential Exposure to Large-ScalePesticide Application. *Journal of Exposure Analysis and Environmental Epidemiology* 12:64–80.

Brown, Aleta and Patrick McCully. 1997. Uniting to Block the Dams. *Multinational Monitor* 18:6–7.

Brown, Phil, Brian Mayer, Steve Zavestoski, Theo Luebke, Josh Mandlebaum, and Sabrina McCormick. 2003. The Politics of Asthma Suffering: Environmental Justice and the Social Movement Transformation of Illness Experience. *Social Science and Medicine* 57(3): 453–464.

Brown, Phil, Sabrina McCormick, Brian Mayer, Steve Zavestoski, Rachel Morello-Frosch, Rebecca Gasior Altman, and Laura Senier. 2006. A Lab of Our Own: Environmental Causation of Breast Cancer and Challenges to the Dominant Epidemiological Paradigm. *Science, Technology, and Human Values* 31(5): 499–536.

Brown, Phil and Edwin Mikkelsen. 1990. *No Safe Place: Toxic Waste, Leukemia, and Community Action.* Berkeley: University of California Press.

Brown, Phil, Stephen Zavestoski, Sabrina McCormick, Joshua Mandelbaum, Theo Luebke, and Meadow Linder. 2001. Gulf of Difference: Disputes over Gulf War–Related Illnesses. *Journal of Health and Social Behavior* 42:235–257.

Brown, Phil, Stephen Zavestoski, Sabrina McCormick, Brian Mayer, Rachel Morello-Frosch, and Rebecca Gasior. 2005. Health Social Movements and Contested Illnesses. In *The New Political Sociology of Science: Institutions, Networks, and Power,* Kelly Moore and Scott Frickel, eds. Madison, WI: University of Wisconsin Press.

Brownlee, Shannon. 2003, January 16. The Perils of Prevention. *New York Times* editorial.

Bruneau, Thomas C. 1986. Brazil: The Catholic Church and the Popular Movement in Nova Iguaçu, 1974–1985. In *Religion and Political Conflict in Latin America,* Daniel H Levine, ed. Chapel Hill, NC: University of North Carolina Press.

Buell, P. 1973. Changing Incidence of Breast Cancer in Japanese-American Women. *Journal of the National Cancer Institute* 51:1479–1483.

Bullard, Robert. 1990. *Dumping in Dixie: Race, Class, and Environmental Quality.* Boulder, CO: Westview Press.

California Breast Cancer Research Program Website. 2003. Genetic and Environmental Modifiers of Breast Cancer Risk. Oakland, CA: California

Breast Cancer Research Program. (http://www.cbcrp.org/research/PageGrant.asp?grant_id=1813).
Campaign for Safe Cosmetics Website. 2005. (http://www.safecosmetics.org).
Cardoso, Ruth Correa Leite. 1992. Popular Movements in the Context of Consolidation of Democracy in Brazil. In *The Making of Social Movements in Latin America: Identity, Strategy, and Democracy,* Arturo Escobar and Sonia Alvarez, eds. Boulder, CO: Westview Press.
Carvalho, Renata. 1999. A Amazônia rumo ao "ciclo da soja" Amazônia Papers No. 2. São Paulo, Brazil: Programa Amazônia, Amigos da Terra. (http://www.amazonia.org.br).
Casamayou, M. H. 2001. *The Politics of Breast Cancer.* Washington, DC: Georgetown University Press.
Castells, Manuel. 2004. *The Power of Identity.* Malden, MA: Blackwell.
Centers for Disease Control and Prevention Website. 2003. *Second National Report on Human Exposures to Carcinogens.* Atlanta, GA: Centers for Disease Control. (http://www.cdc.gov/exposurereport).
Cevallos, Diego. 2006. LATIN AMERICA: Wave of Opposition Hits Hydroelectric Dams. Interpress Service. Rome, Italy: Inter Press Service. (http://www.ipsnews.net/news.asp?idnews=33150).
Chambers, Robert. 1997. *Whose Reality Counts? Putting the First Last.* London: Intermediate Technology Publications, Ltd. (ITDG) Publishing.
Chase, Jacquelyn. 1997. Exodus Revisited: The Politics and Experience of Rural Loss in Central Brazil. *Sociologia ruralis* 39(2):165–185.
Companhia Hidro-Elétrica do São Francisco. 2004. CHESF 2004.
Clapp, Richard W., Deborah T. Forter, and Genevieve K. Howe. 2005, April 1. Harvard Research with Industry Ties Receives $5.9 Million from Defense Breast Cancer Research Program. *Stop the Epidemic: The Newsletter of the Massachusetts Breast Cancer Coalition.* Quincy, MA: Massachusetts Breast Cancer Coalition. (http://mbcc.org/content.php?id=142).
Clarke, Christina A., Sally L. Glasser, Dee W. West, Rochelle R. Ereman, Christina A. Erdmann, Janice M. Barlow, and Margaret R. Wrensch. 2002. Breast Cancer Incidence and Mortality Trends in an Affluent Population: Marin County, California, USA, 1990–1999. *Breast Cancer Research* 4(6):1–7.
Cohen, Jean and Andrew Arato. 1990. *Civil Society and Political Theory.* Cambridge, MA: MIT Press.
Colborn, Theo, Dianne Dumanoski, and Peterson Myers. 1997. *Our Stolen Future.* New York: Penguin Books.
Collins, H. M. 1999. The TEA Set: Tacit Knowledge and Scientific Networks. In *The Science Studies Reader,* Mario Biagioli, ed. New York: Routledge.
Commission of the European Communities. 2000. Communication from the Commission on the Precautionary Principle. Brussels, Belgium: Commission of the European Communities. (http://ec.europa.eu/dgs/health_consumer/library/pub/pub07_en.pdf).

Cone, Marla. 2007, March 4. Public Health Agency Linked to Chemical Industry. *Los Angeles Times.*

Congressionally Directed Medical Research ProgramWebsites. 2003. Fact Sheet: Department of Defense Breast Cancer Research Program. http://cdmrp.army.mil/pubs/factsheets/bcrpfactsheet.htm).

Conklin, Beth. 2002. Shamans versus Pirates in the Amazonian Treasure Chest. *American Anthropologist* 104(4): 1050–1062.

Conrad, Peter. 1992. Medicalization and Social Control. *Annual Review of Sociology* 18:209–232.

———. 1999. A Mirage of Genes. *Sociology of Health and Illness* 21(2): 228–241.

Cooke, Bill and Uma Kothari, eds. 2001. *Participation: The New Tyranny?* New York: Zed Books.

Corburn, Jason. 2005. *Street Science: Community Knowledge and Environmental Health Justice.* Cambridge, MA: MIT Press.

Cornwall, Andrea. 1995. What Is Participatory Research? *Social Science and Medicine* 41(12): 1667–1676.

Cosmetics, Toiletry and Fragrance Association Website. 2003. How Cosmetics Are Regulated. (http://www.ctfa.org/Content/NavigationMenu/Consumer_Information/Cosmetic_Regulation.htm).

da Costa, Ricardo Cunha. 2003. Personal communication.

Coy, Patrick, and Timothy Heeden. 2005. A Stage Model of Social Movement Co-optation: Community Mediation in the United States. *Sociological Quarterly* 46:405–435.

Dant, Tim. 1991. *Knowledge, Ideology and Discourse: A Sociological Perspective.* New York: Routledge.

Davis, Devra Lee. 2002. *When Smoke Ran like Water: Tales of Environmental Deception and the Battle against Pollution.* New York: Basic Books.

Davis, Devra Lee, and H. Leon Bradlow. 1995. Can Environmental Estrogens Cause Breast Cancer? *Scientific American,* Fall:166–172.

Davis, Devra Lee, and Pamela S. Webster. 2002. The Social Context of Science: Cancer and the Environment. *Annals of the American Academy of Political and Social Science* 584(1): 13–34.

Della Porta, Donatella, and Mario Diani. 2006. *Social Movements: An Introduction.* 2nd ed. Malden, MA: Blackwell Publishers.

Department of Health and Human Services. 2003. *Cancer and the Environment: What You Need to Know and What You Can Do.* Washington, DC: National Institute of Health Publications.

DeSombre, Elizabeth R. 2000. *Domestic Sources of International Environmental Policy: Industry, Environmentalists and U.S. Power.* Cambridge, MA: MIT Press.

Dickson, David. 2004. Science and Its Public: The Need for a Third Way. *Social Studies of Science* 30(6): 917-923.

Dixit, A. M., and C. P. Geevan. 2002. A Quantitative Analysis of Plant Use as a Component of EIA: Case of Narmada Sagar Hydroelectric Project in Central India. *Current Science* 79(2): 202–210.

Dorgan, Joanne F., John Brock, Nathaniel Rothman, Larry Needham, and Rosetta Miller. 1999. Serum Organochlorine Pesticides and PCBs and Breast Cancer Risk: Results from a Prospective Analysis. *Cancer Causes and Control* 10:1–11.

Dorsey, Michael. 2004. Political Ecology of Bioprospecting in Amazonian Ecuador: History, Political Economy and Knowledge. In *Contested Nature: Promoting International Biodiversity Conservation with Social Justice in the Twenty-first Century,* Steven R. Brechin, Peter R. Wilshusen, Crystal L. Fortwangler, and Patrick C. West, eds. New York: State University of New York Press.

Drori, Gili, John W. Meyer, Francisco O. Ramirez, and Evan Schofer. 2003. *Science in the Modern World Polity: Institutionalization and Globalization.* Stanford, CA: Stanford University Press.

Dryzek, John S. 2002. *Deliberative Democracy and Beyond.* Oxford: Oxford University Press.

———. 2005. Deliberative Democracy in Divided Societies: Alternatives to Agonism and Analgesia. *Political Theory* 33(2): 218–242.

Dutton, Diane. 1984. The Impact of Public Participation in Biomedical Policy: Evidence from Four Case Studies. In *Citizen Participation in Science Policy,* J. C. Peterson, ed. Amherst: University of Massachusetts Press.

Eckstein, Susan. 1989. *Power and Popular Protest: Latin American Social Movements.* Berkeley: University of California Press.

The Economist. 2003. According to Damming Evidence. July 19.

Egan, Michael. 2007. *Barry Commoner and the Science of Survival: The Remaking of American Environmentalism.* Cambridge, MA: MIT Press.

Eletrobras. 2003. A Contemporary History of Energy Generation. Presentation at the World Social Forum, Porto Alegre. Rio Grande do Sul, Brazil: World Social Forum.

Elster, Jon, and Adam Przeworski. 1998. *Deliberative Democracy.* Cambridge, England: Cambridge University Press.

Environmental Defense. 2004. Controversial Lao Dam Not Suitable for World Bank Support. Press backgrounder. New York: Environmental Defense.

Epstein, Samuel S. 1999. American Cancer Society: The World's Wealthiest "Non-profit" Institution. *International Journal of Health Services* 29(3): 565–78.

———. 2005. Why We Are Still Losing the Winnable Cancer War. *Humanist,* January/February 2005.

Epstein, Steven. 1995. The Construction of Lay Expertise: AIDS Activism and the Forging of Credibility in the Reform of Clinical Trials. *Science, Technology, and Human Values* 20:408–437.

———. 1996. *Impure Science: AIDS, Activism, and the Politics of Knowledge.* Berkeley: University of California Press.

———. 2001. U.S. AIDS Activism and the Question of Epistemological Radicalism. Presentation at the annual meeting of the Society for the Social Study of Science, Cambridge, MA, November.

Escobar, Arturo, and Sonia Alvarez, eds. *The Making of Social Movements in Latin America: Identity, Strategy and Democracy.* Boulder, CO: Westview Press.

Estlund, David. 1997. Beyond Fairness and Deliberation: The Epistemic Dimension of Democratic Authority. In *Deliberative Democracy: Essays on Reason and Politics,* James Bohman and William Rehg, eds. Cambridge, MA: MIT Press.

European Council. 2003. *Innovation Policy: Updating the Union's Approach in the Context of the Lisbon Strategy.* Communication from the Commission to the Council, the European Parliament, the European Economic and Social Committee, and the Committee of the Regions. COM 112. Brussels: CEC.

Evans, Nancy. 2006. *State of the Evidence: What Is the Connection between Breast Cancer and Environment?* San Francisco: Breast Cancer Fund.

Evans, Peter. 1979. Dependent Development: The Alliance of Multinational, State and Local Capital in Brazil. Princeton, NJ: Princeton University Press.

———. 1997. *State-Society Synergy: Government Action and Social Capital in Development.* Berkeley: University of California at Berkeley, International and Area Studies Publications.

———, ed. 2002. *Livable Cities? Urban Struggles for Livelihood and Sustainability.* Berkeley: University of California Press.

Eve, Evaldice, Fracisco A. Arguelles, and Philip Fearnside. 2002. How Well Does Brazil's Environmental Law Work in Practice? Environmental Impact Assessment and the Case of Itapiranga Private Sustainable Logging Plan. *Environmental Management* 26(3): 251–267.

Ezrahi, Yaron. 1990. *Science and the Descent of Icarus: Science and the Transformation of Contemporary Democracy.* Cambridge, MA: Harvard University Press.

Fagin, Dan. 2002a. Breast Cancer Causes Still Elusive, Study: No Clear Link between Pollution, Breast Cancer on Long Island. *Newsday,* August 6.

———. 2002b. Tattered Hopes: A $30 Million Federal Study of Breast Cancer and Pollution on Long Island Has Disappointed Activists and Scientists. *Newsday,* July 28.

Fairhead, James, and Melissa Leach. 2003. *Science, Society and Power: Environmental Knowledge and Policy in West Africa and the Caribbean.* Cambridge, England: Cambridge University Press.

FASE. 2002. *Brazil 2002: The Sustainability That We Want* and *Alternative Sources of Energy.* Rio de Janeiro, Brazil: FASE.

Fearnside, Philip M. 1990. Environmental Destruction in the Brazilian Amazon. In *The Future of Amazonia: Destruction or Sustainable Development?* David Goodman and Anthony Hall, eds., 179–225. London: Macmillan.

———. 1995. Global Warming Response Options in Brazil's Forest Sector: Comparison of Project-Level Costs and Benefits. *Biomass and Bioenergy* 8(5): 309–322.

———. 2002a. Avanca Brasil: Environmental and Social Consequences of Brazil's Planned Infrastructure Project in Amazonia. *Environmental Management* 30(6): 735–747.

———. 2002b. Greenhouse Gas Emissions from a Hydroelectric Reservoir (Brazil's Tucuruí Dam) and the Energy Policy Implications. *Water, Air, and Soil Pollution* 133(1–4): 69–96.

———. 2005. Deforestation in Brazilian Amazonia: History, Rates, and Consequences. *Conservation Biology* 19(3): 680–688.

———. 2006. Dams in the Amazon: Belo Monte and Brazil's Hydroelectric Development of the Xingu River Basin. *Environmental Management* 38(1): 16–27.

Ferguson, Susan J., and Anne S. Kasper. 2000. Living with Breast Cancer. In *Breast Cancer: Society Shapes an Epidemic,* Anne S. Kasper and Susan J. Ferguson, eds. New York: St. Martin's Press.

Fernandes, Adriana. 2005. Brazil's Federal Domestic Debt Load Up 2.3% in Feb to R$ 845bn. (http://www.aebrazil.com/highlights/2005/mar/16/34.htm).

Fischer, Frank. 2000. *Citizens, Experts, and the Environment: The Politics of Local Knowledge.* Durham, NC: Duke University Press.

Fisher, William H. 1994. Megadevelopment, Environmentalism, and Resistance: The Institutional Context of the Kayapo Indigenous Politics in Central Brazil. *Human Organization* 53(3): 220–232.

Fishman, Jennifer. 2000. Assessing Breast Cancer: Risk, Science and Environmental Activism in an "at Risk" Community. In *Ideologies of Breast Cancer: Feminist Perspectives,* Laura K. Potts, ed. New York: St. Martin's Press.

Flint, Adam. 2003. Participatory Action Research and International Solidarity: Towards a Social Scientific Praxis in Support of Social Movements. American Sociological Association presentation, San Francisco.

Foucault, Michel. 1972. *The Archeology of Knowledge.* New York: Pantheon Books.

———. 1980. *Power/Knowledge: Selected Interviews and Other Writings, 1972–1977.* New York: Pantheon Books.

———. 1994. *Ethics: Subjectivity and Truth,* Paul Rabinow, ed. New York: New Press.

Foweraker, Joe. 1995. *Theorizing Social Movements.* London: Pluto Press.

Francis, Paul. 2001. Participatory Development at the World Bank: The Primacy of Process. In *Participation: The New Tyranny?* New York: Zed Books.

Franco, Mirian. 2006. Barragens Do Madeira: A Construção das hidrelétricas e o direito das populações tradicionais que habitam a região. (www.riomadeiravivo.org/noticias/not388.htm).

Fraser, Nancy. 1989. *Unruly Practices: Power, Discourse and Gender in Contemporary Social Theory.* Minneapolis: University of Minnesota Press and Polity Press.

Freire, Paulo. 1970. *Pedagogy of the Oppressed.* New York: Continuum Publishing Co.

Fung, Archon, and Erik Olin Wright. 2002. *Deepening Democracy: Institutional Innovations in Empowered Participatory Governance.* London: Verso.

Gambetta, Diego. 1998, Can We Trust Trust? In *Trust: Making and Breaking Cooperative Relations.* Diego Gambetta, ed., 213–237. New York: Blackwell.

Gammon, Marilie D., Alfred I. Neugut, Regina M. Santella, Susan L. Teitelbaum, Julie A. Britton, Mary B. Terry, Sybil M. Eng, Mary S. Wolff, Steven D. Stellman, Geoffrey C. Kabat, Bruce Levin, H. Leon Bradlow, Maureen Hatch, Jan Beyea, David Camann, Martin Trent, Ruby T. Senie, Gail C. Garbowski, Carla Maffeo, Pat Montalvan, Gertrud S. Berkowitz, Margaret Kemeny, Marc Citron, Freya Schnabel, Allan Schuss, Steven Hajdu, Vincent Vinceguerra, Gwen W. Collmann, and G. Iris Obrams. 2002. The Long Island Breast Cancer Study Project: Description of a Multiinstitutional Collaboration to Identify Environmental Risk Factors for Breast Cancer. *Breast Cancer Research and Treatment* 74(3): 235–254.

Gammon, Marilie D., R. M. Santella, A. I. Neugut, S. M. Eng, S. L. Teitelbaum, J. A. Britton, M. B. Terry, Bruce Levin, S. D. Stellman, G. C. Kabat, Maureen Hatch, Ruby Senie, Gertrud Berkowitz, H. Leon Bradlow, Gail Garbowski, Carla Maffeo, Pat Montalvan, Pat Montalvan, Margaret Kemeny, Marc Citron, Freya Schnabel, Allan Scguss, Steven Hajdu, Vincent Vinceguerra, Nancy Niguidula, Karen Ireland and Regina M. Santella. 2002. Environmental Toxins and Breast Cancer on Long Island. I. Polycyclic Aromatic Hydrocarbon DNA Adducts. *Cancer Epidemiology and Biomarkers Prevention* 11:677–685.

Gamson, William. 1990 [1975]. *The Strategy of Social Protest.* Homewood, IL: Dorsey Press.

———. 1992. The Social Psychology of Collective Action. In *Frontiers in Social Movement Theory,* Aldon Morris and Carol Mueller, eds. New Haven, CT: Yale University Press.

Gamson, William A. and David S. Meyer. 1996. Framing Political Opportunity. In *Comparative Perspectives on Social Movements: Political Opportunities, Mobilizing Structures, and Cultural Framings,* Doug McAdam, John D. McCarthy, and Mayer N. Zald, eds. Cambridge: Cambridge University Press.

Gastil, John, and Peter Levine. 2005. *The Deliberative Democracy Handbook: Strategies for Effective Civic Engagement in the Twenty-First Century.* San Francisco: Jossey-Bass.

Gaventa, John. 1993. The Powerful, the Powerless and the Experts: Knowledge Struggles in Information Age. In *Participatory Research in North America,* Peter Park, Budd Hall, and Ted Jackson, eds. Amherst, MA: Bergin and Hadley.

Gibbs, Lois. 1982. *Love Canal: My Story.* Albany: State University of New York Press.

Gieryn, Thomas. 1983. Boundary Work in Professional Ideology of Scientists. *American Sociological Review* 48:781–795.

Giugni, Marco G. 1998. Was It Worth the Effort? The Outcomes and Consequences of Social Movements. *Annual Review of Sociology* 24: 371–393.

Giugni, Marco, Doug McAdam, and Charles Tilly. 1999. *How Social Movements Matter.* Minneapolis: University of Minnesota Press.

Glasson, John, and Andrew Chadwick. 2005. *Introduction to Environmental Impact Assessment.* Oxford, England:Taylor and Francis.

Goldenberg, Jose, Suani T. Coelho, and Fernando Rei. 2002. Brazilian Energy Matrix and Sustainable Development. *Energy for Sustainable Development* 5(4): 55–59.

Goldman, Lynn. 1998. Linking Research and Policy in Order to Assure Children's Environmental Health. *Environmental Health Perspectives* 106(3): 857–862.

Goldman, Michael. 2001. *Imperial Nature: The World Bank and Struggles for Social Justice in the Age of Globalization.* New Haven, CT: Yale University Press.

Gore, Al. 2006. *Earth in the Balance: Ecology and the Human Spirit.* London: James and James/Earthscan.

Gottlieb, Robert. 2005. *Forcing the Spring: The Transformation of the American Environmental Movement.* Washington, DC: Island Press.

Goulet, Denis. 2005. Global Governance, Dam Conflicts, and Participation. *Human Rights Quarterly* 27(3): 881–907.

Gremmen, Bart and Henk van de Belt. 2000. The Precautionary Principle and Pesticides. *Journal of Agricultural and Environmental Ethics* 12:197–205.

Griffith, Jack, R. C. Duncan, W. B. Riggan, and A. C. Pellom. 1989. Cancer Mortality in U.S. Counties with Hazardous Waste Sites and Ground Water Pollution. *Archives of Environmental Health* 44(2): 69–74.

Guest, Greg, and Eric C. Jones. 2005. Globalization, Health and the Environment: An Introduction. In *Globalization, Health and the Environment: An Integrated Perspective.* Lanham, MD: Rowman and Littlefield.

Guidry, John A., Michael D. Kennedy, and Mayer N. Zald. 2000. *Globalizations and Social Movements: Culture, Power, and the Transnational Public Sphere.* Ann Arbor: University of Michigan Press.

Guston, David. 1999. Stabilizing the Boundary between U.S. Politics and Science: The Role of the Office of Technology Transfer as a Boundary Organization. *Social Studies of Science* 29(1): 87–112.

Gutman, Amy, and Dennis Thompson. 2004. *Why Deliberative Democracy?* Princeton, NJ: Princeton University Press.

Guttes, S., K. Failing, K. Neumann, J. Kleinstein, S. Georgii, and H. Brunn. 1998. Chlororganic Pesticides and Polychlorinated Biphenyls in Breast Tissue of Women with Benign and Malignant Breast Disease. *Archives of Environmental Contamination and Toxicology* 35(1): 140–147.

Haas, Peter. 1992. Introduction: Epistemic Communities and International Policy Coordination. *International Organization* 46(1): 1–37.

Habermas, Jürgen. 1970. *Toward a Rational Society: Student Protest, Science, and Politics.* Translated by Jeremy J. Shapiro. Boston: Beacon Press.

———. 1972. Science and Technology as Ideology. In *Sociology of Science: Selected Readings.* Barry Barnes, ed. Toronto: Penguin Books.

———. 1984. *The Theory of Communicative Action.* Boston: Beacon Press.

———. 1989. *The Structural Transformation of the Public Sphere: An Inquiry into a Category of Bourgeois Society.* Cambridge, MA: MIT Press.

Hansen, Johnni. 1999. Breast Cancer Risk among Relatively Young Women Employed in Solvent-Using Industries. *American Journal of Industrial Medicine* 36:43–47.

Haraway, Donna. 1988. Situated Knowledges: The Science Question in Feminism and the Privilege of Partial Perspective. *Feminist Studies* 14:575–599.

Harding, Sandra. 1998a. Gender, Development, and Post-Enlightenment Philosophies of Science. *Hypatia* 13(3): 146–167.

———. 1998b. *Is Science Multicultural?* Bloomington: Indiana University Press.

———. 2004. *The Feminist Standpoint Theory Reader.* New York: Routledge.

Harvey, Philip. W., and Philippa Darbre. 2004. Endocrine Disrupters and Human Health: Could Oestrogenic Chemicals in Body Care Cosmetics Adversely Affect Breast Cancer Incidence in Women? A Review of Evidence and Call for Further Research. *Journal of Applied Toxicology* 24(3): 167–176.

Heller, Patrick. 2001a. Moving the State: The Politics of Decentralization in Kerala, South Africa, and Porto Alegre. *Politics and Society* 29(1): 131–163.

———. 2001b. Civil Society and Democracy: The Antinomies of the South African Case. Interuniversity Seminar Series on Comparative Development. Presented at the annual meeting of the American Sociological Association. Watson Institute, Brown University.

———. 2002. Globalization Lecture. Brown University.

Hess, David. 1995. *Science and Technology in a Multicultural World.* New York: Columbia University Press.

———. 1997. *Science Studies: An Advanced Introduction.* New York: New York University Press.

———. 2005. Technology- and Product-Oriented Movements: Approximating Social Movement Studies and Science and Technology Studies. *Science, Technology, and Human Values* 20:515–535.

———. 2007. *Alternative Pathways in Science and Industry: Activism, Innovation and the Environment in an Era of Globalization.* Cambridge, MA: MIT Press.

Hickey, Sam. 2002. Transnational NGOS and Participatory Forms of Rights-Based Development: Converging with the Local Politics of Citizenship in Cameroon. *Journal of International Development* 14:841–857.

Hokanson, R., W. Hanneman, M. Hennessey, K. C. Donnelly, T. McDonald, R. Chowdhary, and D. L. Busbee. 2006. DEHP, Bis(2)-Ethylhexyl Phthalate, Alters Gene Expression in Human Cells: Possible Correlation with Initiation of Fetal Developmental Abnormalities. *Human and Experimental Toxicology* 25(12): 687–695.

House Committee on Science and Technology. 2002. Congressional Hearing, Port Jefferson Hall, Long Island. (http://gop.science.house.gov/hearings/ets02/jun22/pace.htm).

House of Representatives. 2003a. www.house.gov/lantos/html_files/womens_protecting.html.

———. 2003b. Study of Elevated Breast Cancer Rates in Long Island: Public Law 103-43, June 10, 1993.

Hoyer, A. P., T. Jorgensen, J. W. Brock, and P. Grandjean. 2000. Organochlorine exposure and breast cancer survival. *Journal of Clinical Epidemiology* 53(3): 323–30.

Hoyer, A. P., T. Jorgensen, J. W. Brock, and P. Grandjean. 2000. Organochlorine Exposure and Breast Cancer Survival. *Journal of Clinical Epidemiology* 53(3): 323–330.

Hsiao, H.-H. M., and H.-J. Liu. 2002. Collective Action toward a Sustainable City: Citizens'

Movements and Environmental Politics in Taipei. In *Livable Cities? Urban Struggles for Livelihood and Sustainability,* Peter Evans, ed., 67–94. Berkeley: University of California Press.

Hubbard, Ruth. 1990. *The Politics of Women's Biology.* New Brunswick, NJ: Rutgers University Press.

Hunter, David J., Susan Hankinson, Francine Laden, Graham Colditz, JoAnne Manson, Walter Willett, Frank Speizer, and Mary S. Wolff. 1997. Plasma Organochlorine Levels and the Risk of Breast Cancer. *New England Journal of Medicine* 337:1253–1258.

Hunter, David J., and Karl T. Kelsey. 1993. Pesticide Residues and Breast Cancer: The Harvest of the Silent Spring? *Journal of the National Cancer Institute* 85(8): 598–599.

Hutchinson, Thomas, Rick Brown, Kristin Brugger, Pamela Campbell, Martin Holt, Rienhard Lange, Peter MaCahon, Lisa Tattersfield, and Roger van Egmond. 2000. Ecological Risk Assessment of Endocrine Disruptors. *Environmental Health Perspectives* 108(11): 1007–1014.

Israel, B. A., Amy J. Schulz, Edith A. Parker, and Adam B. Becker. 1998. Review of Community-Based Research: Assessing Partnership Approaches to Improve Public Health. *Annual Review of Public Health* 19:173–202.

Jaquette, J. S. 1994. *The Women's Movement in Latin America: Participation and Democracy.* Boulder, CO: Westview Press.

Jasanoff, Sheila. 1990. *The Fifth Branch: Science Advisers as Policymakers.* Cambridge, MA: Harvard University Press.

———. 1996. Compelling Knowledge in Public Decisions. In L.A. Brooks and S. VanDeveer, eds. *Saving the Seas: Values, Scientists, and International Governance.* College Park, MD: Maryland Sea Grant, pp. 229–252.

———. 2006. Biotechnology and Empire: The Global Power of Seeds and Science. *Osiris* 21:273–292.

Judge, Paramjit. 1997. Response to Dams and Displacement in Two Indian States. *Asian Survey* 37(9): 840–851.

Kant, Ashima K., Arthur Schatzkin, Barry I. Graubard, and Catherine Schairer. 2000. A Prospective Study of Diet Quality and Mortality in Women. *Journal of the American Medical Association* 283(16): 2109–2115.

Kasper, Anne S., and Susan J. Ferguson, eds. 2000. *Breast Cancer: Society Shapes an Epidemic.* New York: St. Martin's Press.

Khagram, Sanjeev. 2004. *Dams and Development: Transnational Struggles for Water and Power.* Ithaca, NY: Cornell University Press.

Khagram, Sanjeev, James Riker, and Kathryn Sikkink. 2000. *Restructuring World Politics: Transnational Social Movements, Networks, and Norms.* Minneapolis: University of Minnesota Press.

King, Samantha. 2006. *Pink Ribbon, Inc.: Breast Cancer and the Politics of Philanthropy.* Minneapolis: University of Minnesota Press.

Klawiter, Maren. 1999. Racing for the Cure, Walking Women, and Toxic Touring: Mapping Cultures of Action within the Bay Area Terrain of Breast Cancer. *Social Problems* 29(3): 104–127.

———. 2000. Racing for the Cure, Walking Women and Toxic Touring: Mapping Cultures of Action within the Bay Area Terrain of Breast Cancer. In *Ideologies of Breast Cancer: Feminist Perspectives,* L. K. Potts, ed., 63–97. New York: St. Martin's Press.

———. 2001. Breast Cancer Activism in the U.S.: Diverse Perspectives. Presented at the annual meeting of the Society for the Social Study of Science, Cambridge, MA.

Knorr-Cetina, K. 1999. *Epistemic Cultures: How the Sciences Make Knowledge.* Cambridge, MA: Harvard University Press.

Koch, T. 2005. The Challenge of Terri Schiavo: Lessons for Bioethics. *Journal of Medical Ethics* 31:376–378.

Kolata, Gina. 2002. The Epidemic That Wasn't. *New York Times,* August 29.

Kolker, Emily S. 2004. Framing as a Cultural Resource in Health Social Movements: Funding Breast Cancer Activism in the U.S., 1990–1993. *Social Science and Medicine* 26(6): 820–844.

Kriebel, David, Joel Tickner, Paul Epstein, John Lemons, Richard Levins, Edward L. Loechler, Margaret Quinn, Ruthann Rudel, Ted Schettler, and Michael Soto. 2001. The Precautionary Principle in Environmental Science. *Environmental Health Perspectives* 109:871–875.

Krieger, Nancy. 1989. Exposure, Susceptibility, and Breast Cancer Risk: A Hypothesis Regarding Exogenous Carcinogens, Breast Tissue Development, and Social Gradients, Including Black/White Differences, in Breast Cancer Incidence. *Breast Cancer Research and Treatment* 13(3): 205–223.

———. 2003. Social Production of Disease/Political Economy of Health. *American Journal of Public Health* 93(2): 194–199.

Krieger, Nancy, and Sally Zierler. 1995. What Explains the Public's Health: A Call for Epidemiologic Theory. *Epidemiology* 7:107–109.

Krimsky, Sheldon. 1984. *The Social History of the Recombinant DNA Controversy.* Cambridge, MA: MIT Press.

———. 2000. *Hormonal Chaos: The Scientific and Social Origins of the Endocrine Disrupter Hypothesis.* Baltimore: Johns Hopkins University Press.

———. 2003. Science on Trial. *GeneWatch* 16(5): 3–6.

Kuhn, Thomas. 1962. *The Social Structure of Scientific Revolutions.* Chicago: University of Chicago Press.

Labreche F.P., and Goldberg. 1997. Exposure to Organic Solvents and Breast Cancer in Women: A Hypothesis. *American Journal of Industrial Medicine* 32(1): 1–14.

Landigran, P. J., Joy E. Carlson, Cynthia F. Bearer, Joan Spyker Cranmer, Robert D. Bullard, Ruth A. Etzel, John Groopman, John A. McLachlan, Frederica P. Perera, J. Routt Reigart, Leslie Robinson, Lawrence Schell, and William A. Suk. 1998. Children's Health and the Environment: A New Agenda for Prevention Research. *Environmental Health Perspectives* 106(supplement 3): 787–794.

Lane, Robert E. 1966. The Decline of Politics and Ideology in a Knowledge Society. *American Sociological Review* 31:649–662.

Langer, Erick Detlef, and Elena Muñoz. 2003. *Contemporary Indigenous Movements in Latin America.* Wilmington, DE: SR Books.

LaRovere, Emilio L., and Fernando Eduardo Mendes. 2000. *Brazil Case Study: Tucuruí Dam and Amazon/Tocantins River Basin.* Cape Town, South Africa: World Commission on Dams.

Latour, Bruno. 1993. *We Have Never Been Modern.* Cambridge, MA: Harvard University Press.

Laurance, William F., Mark A. Cochrane, Scott Bergen, Philip M Fearnside, Patricia Delamônica, Christopher Barber, Sammy D'Angelo, andTito Fernandes. 2001. The Future of the Brazilian Amazon. *Science* 291(5503): 438–439.

Layder, Derek. 1994. *Understanding Social Theory.* London: University of Leicester Press.

Lemos, Maria Carmen, and Joao Lucio Farias de Oliveira. 2004. Can Water Reform Survive Politics? Institutional Change and River Basin Management in Ceara, Northeast Brazil. *World Development* 32(12): 2121–2137.

Lemos, Maria and J. Timmons Roberts. 2008. Environmental policy-making networks and the future of the Amazon. *Philosophical Transactions of the Royal Society* 363(1498): 1897–1902.

Leopold, Ellen. 2001. Shopping for the Cure. *Breast Cancer Action Newsletter* 63. (http://bcaction.org/index.php?page=newsletter-63a).

Lewis, Carol. 1998. Clearing Up Cosmetic Confusion. U.S. Food and Drug Administration. *FDA Consumer* May/June 1998. (http://www.pueblo.gsa.gov/cic_text/health/cosmetic-confusion/398_cosm.html).

Leydesdorff, Loet, and Janelle Ward. 2005. Science Shops: A Kaleidoscope of Science-Society Collaborations in Europe. *Public Understanding of Science* 14(4): 353–372.

Lichtenstein, P., N. Holm, P. K. Verkasalo, A. Iliadou, J. Kaprio, M. Koskenvuo, E. Pukkala, A. Skytthe, and K. Hemminki. 2000. Environmental and Heritable Factors in the Causation of Cancer—Analyses of Cohorts of Twins from Sweden, Denmark, and Finland. *New England Journal of Medicine* 343:78–85.

Longino, Helene E. 1990. *Science as Social Knowledge: Values and Objectivity in Scientific Inquiry.* Princeton, NJ: Princeton University Press.

Long Island Breast Cancer Study Project. 2002. Summit proceedings.

Lourde, Audre. 1980. *The Cancer Journals.* Minneapolis: Aunt Lute Books.

Luján, José Luis, and Oliver Todt. 2007. Precaution in Public: The Social Perception of the Role of Science and Values in Policy Making. *Public Understanding of Science* 16(1): 97–109.

Luxner, Larry. 1991. Watt a Dam! *Americas* 43:2–3.

MacDonald, Gordon J., Daniel L. Nielson, and Marc A. Stern, eds. 1997. *Latin American Environmental Policy in International Perspective.* Boulder, CO:Westview Press.

MacMahon, B. 1994. Cigarette Smoking and Risk of Fatal Breast Cancer. *American Journal of Epidemiology* 139:1001–1007.

Macnaghten, Phil, Matthew B. Kearnes, and Brian Wynne. 2005. Nanotechnology, Governance, and Public Deliberation: What Role for the Social Sciences? *Science Communication* 27(2): 268–291.

Mannheim, Karl. 1952. *Essays on the Sociology of Knowledge.* New York: Oxford University Press.

Martins, S. J., and R. C. Menezes. 1994. Nutritional Status of Parakana Indian Children from Birth through 5 Years of Age, Eastern Amazonia. *Revista de saude publica* 28(1): 1–8.

Massachusetts Breast Cancer Coalition. 2006. MBCC History and Accomplishments. Quincy, MA. (http://mbcc.org/content.php?id=17).

Masson, Stéphane, and Alain Tremblay. 2003. Effects of Intensive Fishing on the Structure of Zooplankton Communities and Mercury Levels. *Science of the Total Environment* 304(1–3): 377–390.

McAdam, Doug. 1982. *Political Process and the Development of Black Insurgency*. Chicago: University of Chicago Press.

McAdam, Doug, Sidney Tarrow, and Charles Tilly. 2001. *Dynamics of Contention*. New York: Cambridge University Press.

McCarthy, John D., and Mayer N. Zald. 1973. *The Trend of Social Movements in America: Professionalization and Resource Mobilization*. Morristown, NJ: General Learning.

———. 1977. Resource Mobilization and Social Movements: A Partial Theory. *American Journal of Sociology* 82:1212–1241.

McCormick, Sabrina. 2001. The Personal Is Scientific, the Scientific Is Political: The Environmental Breast Cancer Movement. Master's thesis, Brown University, Providence, RI.

———. 2008. From "Politico-scientists" to Democratizing Science Movements: The Changing Climate of Citizens and Science. *American Sociological Association Presentation*. Boston, MA.

McCormick, Sabrina, and Lori Baralt. 2007. Movement Success or Cooptation? The Breast Cancer Movement. Unpublished manuscript.

———. 2008. Public Participation in Environmental Breast Cancer Research: A Progress Report. American Academy of the Sciences. Boston, MA.

McCormick, Sabrina, Phil Brown, and Steve Zavestoski. 2003. The Personal Is Scientific, the Scientific Is Political: The Environmental Breast Cancer Movement. *Sociological Forum* 18(4): 545–576.

McCormick, Sabrina, Ruth Polk, Phil Brown, and Julia Brody. 2004. Public Involvement in Breast Cancer Research: An Analysis and Prototype. *International Journal of Health Services* 34(4): 625–646.

McCright, Aaron, and Riley Dunlap. 2003. Defeating Kyoto: The Conservative Movement's Impact on U.S. Climate Change. *Social Problems* 50(3): 358–373.

McCully, Patrick. 2001. *Silenced Rivers: The Ecology and Politics of Large Dams*. London: Zed Books.

McDonald, Mark D. 1993. Dams, Displacement, and Development: A Resistance Movement in Southern Brazil. In *Defense of Livelihood: Comparative Studies on Environmental Action,* H. Rangan, ed. West Hartford, CT: Kumarian Press.

McTaggart, Robin. 1991. *Participatory Action Research: International Contexts and Consequences*. New York: SUNY Press.

Meehan, Elizabeth. 1996. Democracy Unbound. *Reconstituting Politics*. Belfast, Ireland: Democratic Dialogue.

Melucci, Alberto. 1996. *Challenging Codes: Collective Action in the Information Age*. New York: Cambridge University Press.

———. 1997 [1985]. The Symbolic Challenge of Contemporary Movements. In *Social Movements: Perspectives and Issues*. Steven M. Buechler and F. Kurt Cylke, eds. La Honda, CA: Mountain View Press.

Meredith, Larry. 2002. Monthly Breast Cancer Status Reports. Marin County Department of Health and Human Services.

Merton, Robert King. 1996. *On Social Structure and Science*. Chicago: University of Chicago Press.

Meyer, David S. 1993. Institutionalizing Dissent: The United States Structure of Political Opportunity and the End of the Nuclear Freeze Movement. *Sociological Forum* 8(2): 157–179.

———. 2002. Opportunities and Identities: Bridge Building in the Study of Social Movements. In *Social Movements: Identity, Culture and the State*, David S. Meyer, Nancy Whittier, and Belinda Robnett, eds. Oxford: Oxford University Press.

Meyer, David S., and Sidney Tarrow. 1998. *The Social Movement Society: Contentious Politics for a New Century*. Lanham, MD: Rowman and Littlefield.

Meyer, John W. 1987. *The World Polity and the Authority of the Nation-State. Institutional Structure: Constituting State, Society and the Individual*. Beverly Hills, CA: Sage.

———. 2000. Globalization: Sources and Effects on National States and Societies. *International Sociology* 15(2): 233–248.

Meyer, John W., John Boli, George M. Thomas, and Francisco M. Ramirez. 1997. World Society and the Nation-State. *American Journal of Sociology* 103(1): 144–181.

Miller, Clark. 2001. Hybrid Management: Boundary Organizations, Science Policy, and Environmental Governance in the Climate Regime. *Science, Technology & Human Values* 26(4): 478–500.

Miller, Jon D. 2004. Public Understanding of, and Attitudes toward, Scientific Research: What We Know and What We Need to Know. *Public Understanding of Science* 13(3): 273–294.

Minkler, Meredith. 2000. Using Participatory Action Research to Build Healthy Communities. *Public Health Reports* 115(2–3): 191–197.

Minkler, Meredith and Nina Wallerstein, eds. Budd Hall, foreword. 2002. *Community-Based Participatory Research for Health*. New York: Jossey-Bass.

Mishler, Eliot. 1981. Viewpoint: Critical Perspectives on the Biomedical Model. In *Social Contexts of Health, Illness, and Patient Care*, Eliot Mishler, ed., 1–19. Cambridge, MA: Cambridge University Press.

Mitra, Amal, Fazlay S. Faurque, and Amanda Avis. 2004. Breast Cancer and Environmental Risks: Where Is the link? *Journal of Environmental Health* 66(7): 24–31.

Moore, Kelly. 1996. Organizing Integrity: American Science and the Creation of Public Interest Organizations, 1955–1975. *AJS* 101 (6): 1592–1627.

———. 2008. *Disrupting Science: Social Movements, American Scientists, and the Politics of the Military, 1945–1975*. Princeton, NJ: Princeton University Press.

Morello-Frosch, Rachel, Steve Zavestoski, Phil Brown, Rebecca Gasior Altman, Sabrina McCormick, and Brian Mayer. 2005. Embodied Health Movements: Responses to a "Scientized" World. In *The New Political Sociology of Science: Institutions, Networks, and Power,* Kelly Moore and Scott Frickel, eds. Madison: University of Wisconsin Press.

Movimento dos Atingidos por Barragens. 2000. A Letter to the Inter-American Development Bank. MAB: Sao Paulo, Brazil.

———. 2001a. *Movimento dos Atingidos por Barragens: Manual dos Atingidos*. Brasilia. (http://www.prodema.ufc.br/dissertacoes/156.pdf).

———. 2001b. Letter to Alcoa. MAB: Sao Paulo, Brazil.

———. 2002. Solicitation of Independent Inspection Mechanism Re: Cana Brava Hydroelectric Project (IDB loan BR-0304).

———. 2003. *A Organização do Movimento dos Atingidos por Barragens*. Caderno no 5. Brasilia.

———. 2005. A Crise do Modelo Energetico – A Construir Um Outro Modelo e Possivel. MAB: Brasilia.

Murphree, David, Stuart Wright, and Helen Rose Ebaugh. 1996. Toxic Waste Siting and Community Resistance: How Cooptation of Local Citizen Opposition Failed. *Sociological Perspectives* 39(4): 447–463.

Najem, G. R., and T. W. Greer. 1985. Female Reproductive Organs and Breast Cancer Mortality in New Jersey Counties and the Relationship with Certain Environmental Variables. *Preventive Medicine* 14:620–635.

National Cancer Institute Website. 1998. *Charting the Course: Priorities for Breast Cancer Research*. (http://prg.nci.nih.gov/breast/finalreport.html-NIEHS).

———. 2003. *Long Island Breast Cancer Study Project Funding*. (http://epi.grants.cancer.gov/LIBCSP/).

National Environmental Trust, Breast Cancer Action, and Breast Cancer Fund. 2004. SB 484 General Fact Sheet. San Francisco, CA.

National Institute of Environmental Health Sciences. 2002. (http://grants.nih.gov/grants/guide/rfa-files/RFA-ES-03-001.html).

———. 2003. http://www.niehs.nih.gov/od/tnback.htm.

National Research Council. 1972. *The Effects on Populations of Exposure to Low Levels of Ionizing Radiation: Report to the Advisory Committee on the Biological Effects of Ionizing Radiation*. P 14-53. Washington, DC: U.S. Government Printing Office.

Nelson, Lynn Hankinson. 1990. *Who Knows: From Quine to a Feminist Empiricism*. Philadelphia: Temple University Press.

Niskanen, William A. 2001. Bringing Power to Knowledge: Choosing Policies to Use Decentralized Knowledge. In *Knowledge and Politics,* R. Viale, ed. New York: O'Donnell, Physica-Verlag.

Northridge, Mary E., Joanne Yankura, Patrick L. Kinney, Regina M. Santelle, Regina M. Santella, Peggy Shepard, Ynolde Riojas, Maneesha Aggarwal, Paul Strickland, and the Earth Crew. 1999. Diesel Exhaust among Adolescents in Harlem: A Community-Driven Study. *American Journal of Public Health* 89:998–1002.

Oberschall, Anthony. 1973. *Social Conflict and Social Movements.* Englewood Cliffs, NJ: Prentice-Hall.

O'Donnell, Guillermo, A., Philippe C. Schmitter, and Laurence Whitehead. 1986. *Transitions from Authoritarian Rule: Prospects for Democracy.* Baltimore: Johns Hopkins University Press.

O'Fallon, L. R., Frederick L. Tyson, and Allen Dearry. 2000. Improving Public Health through Community-Based Participatory Research and Outreach. *Environmental Epidemiology and Toxicology* 2:201–209.

O'Rourke, Dara. 2002. Community-Driven Regulation: Toward an Improved Model of Environmental Regulation in Vietnam. In *Livable Cities? Urban Struggles for Livelihood and Sustainability.* Peter Evans, ed., 95–131. Berkeley: University of California Press.

Ortiz, Lucia Schild. 2002. *Fontes alternativas de energia e eficiencia energetica: Opcao para uma politica energetica sustentavel no Brasil.* Campo Grande, Minas Gerais: Ecologia e Açao.

Pain, Rachel, and Peter Francis. 2003. Reflections on Participatory Research. *Area* 35(1): 46–54.

Pellow, David. 1999. Framing Emerging Environmental Movement Tactics: Mobilizing Consensus, Demobilizing Conflict. *Sociological Forum* 14(4): 659–683.

Phipps, Alison. 2006. I Can't Do with Whinging Women! Feminism and the Habitus of "Women in Science" Activists. *Women's Studies International Forum* 29(2): 125–135.

Pickvance, Katy. 1997. Social Movements in Hungary and Russia: The Case of Environmental Movements. *European Sociological Review* 13(1): 35–54.

Pimbert, Michel P., and Tom Wakeford. 2001. *Deliberative Democracy and Citizen Empowerment. Participatory Learning and Action Notes 40.* London: IIED.

Pinto, L. Flavio. 2002. *Hidreletricas na Amazonia: Predestinacao fatalidade ou engodo?* Belém, Pará: Edição Jornal Pessoal.

Polanyi, Karl. 1977. *The Great Transformation.* Boston: Beacon Press.

Prevention First. 2003. http://www.preventionfirstcoalition.org/AboutUS/AboutUs.html. Downloaded December 4, 2003.

Quigley, Diane. 2006. Perspective: A Review of Improved Ethical Practices in Environmental and Public Health Research; Case Examples from Native Communities. *Health Education and Behavior* 33(2): 130–147.

Quinn, Andrew. 2002. California Urges Study of Alarming Breast Cancer Rates. *Reuters Health,* October 24. (http://www.oncolink.org/resources/article.cfm?c=3&s=8&ss=23&id=8992&month=10&year=2002).

Raffensberger, Carolyn, and Joel Tickner. 1999. *Protecting Public Health and the Environment: Implementing the Precautionary Principle*. Washington, DC: Island Press.

Reiss, Joan Reinhardt, and Andrea Ravinett Martin. 2000. *Breast Cancer 2000: An Update on the Facts, Figures and Issues*. San Francisco: Breast Cancer Fund.

Reuters Website. 2007. Brazil to Split Up Its Environmental Agency. April 26. (http://www.planetark.com/dailynewsstory.cfm/newsid/41565/story.htm).

Reyes, Linda. 1993. Ionizing Radiation: Part Two—Mammography Screening. *Newsletter* 16.

Reynolds, Peggy, Julie Von Behren, Robert B. Gunier, Debbie E. Goldberg, Andrew Hertz, and Martha E. Harnly. 2002. Childhood Cancer and Agricultural Pesticide Use: An Ecologic Study in California. *Environmental Health Perspectives* 110(3): 319–324.

Rezende, Leonardo Pereira. 2007. Avanços e contradições do licenciamento ambiental de barragens hidrelétricas. Brazil: Forum.

Robbins, Anthony S., Sonia Brescianni, and Jennifer Kelsey. 1997. Regional Differences in Known Risk Factors and the Higher Incidence of Breast Cancer in San Francisco. *Journal of the National Cancer Institute* 89:960–964.

Roberts, J. Timmons, and Nikki Demetria Thanos. 2003. *Trouble in Paradise: Globalization and Environmental Crisis in Latin America*. New York: Routledge.

Rose, Geoffrey. 1985. Sick Individuals and Sick Populations. *International Journal of Epidemiology* 4:32–38.

Rosenberg, Dorothy Goldin. 2000. "Toward Indigenous Wholeness: Feminist Praxis in Transformative Learning on Health and the Environment. In *Indigenous Knowledges in Global Contexts,* G. J. Sefa Dei, Budd L. Hall, and D. G. Rosenberg, eds. Toronto: University of Toronto Press.

Rothman, Franklin Daniel. 1993. Political Process and Peasant Opposition to Large Hydroelectric Dams: The Case of the Rio Uruguai Movement in Southern Brazil, 1979 to 1992." Ph.D. diss., University of Wisconsin–Madison.

———. 2001. A Comparative Study of Dam-Resistance Campaigns and Environmental Policy in Brazil. *Journal of Environment and Development* 10:317–344.

———, ed. 2008. Vidas alagadas: Conflitos socioambientais, licenciamentoe barragens. Viçosa: Editora UFV.

Rothman, Franklin D., and Pamela Oliver. 1999. From Local to Global: The Brazilian Anti-dam Movement in Southern Brazil, 1979–1992. *Mobilization* 4:41–57.

Rudel, Ruthann A., J. Brody, Julia D. Spengler, John D., Jose Vellarino, Paul W. Geno, Gang Sun, and Alice Yau. 2001. Identification of Selected Hormonally Active Agents and Animal Mammary Carcinogens in Commercial and Residential Air and Dust Samples. *Journal of the Air and Water Management Association* 51:499–513.

Rudel, Ruthann A., D. E. Camann, J. D. Spengler, L. R. Korn, and J. G. Brody. 2003. Phthalates, Alkylphenols, Pesticides, Polybrominated Diphenyl Ethers, and Other Endocrine-Disrupting Compounds in Indoor Air and Dust. *Environmental Science and Technology* 37(20): 4543–4553.

Rudel, Ruthann A., Kathleen R. Attfield, Jessica N. Schifano, and Julia Green Brody. 2007. Chemicals Causing Mammary Gland Tumors in Animals Signal New Directions for Epidemiology, Chemicals Testing, and Risk Assessment for Breast Cancer Prevention. *Cancer Supplement* 109(12): 2635–2665.

Rueschemeyer, Dietrich, and Theda Skocpol. 1996. *States, Social Knowledge, and the Origins of Modern Social Policies*. Princeton, NJ: Princeton University Press.

Russell, M., and M. Gruber. 1987. Risk Assessment in Environmental Policy-Making. *Science* 236(4799): 286–290.

Ruzek, Sheryl Burt. 1980. Medical Response to Women's Health Activities: Conflict, Accommodation and Cooptation. In *Research in the Sociology of Health Care*, pp.325–54, J. A. Roth, ed. Stamford, CT: JAI Press.

Ruzek, Sheryl Burt, Virginia L. Olesen, and Adele Clarke, eds. 1997. *Women's Health: Complexities and Differences*. Columbus: Ohio State University Press.

Ryfe, David M. 2005. Does Deliberative Democracy Work? *Annual Review of Political Science* 8:49–71.

Safe, Stephen. 1997. Xenoestrogens and Breast Cancer. *New England Journal of Medicine* 337:1303–1304.

Santos, Leinad Ayer O., Lúcia M. M. de Andrade, and Robin Wright, eds. 1990. *Hydroelectric Dams on Brazil's Xingu River and Indigenous Peoples*. Cambridge, MA: Cultural Survival.

Schaeffer, Roberto. 1990. Hydroelectric Dams and Nuclear Power. In *Hydroelectric Dams on Brazil's Xingu River and Indigenous Peoples*, Leinad Ayer O. Santos, Lúcia M. M. de Andrade, and Robin Wright, eds. Cambridge, MA: Cultural Survival.

Schattschneider, E. E. 1960. *The Semisovereign People*. New York: Holt, Rinehart and Winston.

Schell, L. M., and A. M. Tarbell. 1998. A Partnership Study of PCBs and the Health of Mohawk Youth: Lessons from Our Past and Guidelines for Our Future. *Environmental Health Perspectives* 106:833–840.

Schettler, T., G. Solomon, M. Valenti, and A. Huddle. 2000. *Generations at Risk: Reproductive Health and the Environment*. Cambridge, MA: MIT Press.

Schibeci, Renato, and Jeffrey Harwood. 2007. Stimulating Authentic Community Involvement in Biotechnology Policy in Australia. *Public Understanding of Science* 16(2): 245–255.

Scott, James. 1998. *Seeing like a State.* New Haven, CT: Yale University Press.

Seiler, Lauren H., and Gene F. Summers. 1979. Corporate Involvement in Community Affairs. *Sociological Quarterly* 20:375–386.

Selznick, Philip. 1953. Foundations of the Theory of Organization. *American Sociological Review* 13(1):25–35.

Shiva, Vandana. 1989. *Staying Alive: Women, Ecology and Development.* London: Zed Books.

Silent Spring Institute. 1998a. *The Cape Cod Breast Cancer and Environment Study: Results of the First Three Years of Study.* Newton, MA: Silent Spring Institute.

———. 1998b. *Findings of the Cape Cod Breast Cancer and Environment Study.* Newton, MA: Silent Spring Institute.

Silveira, José Paulo, William F. Laurance, Philip M. Fearnside, Mark A. Cochrane, Sammya D'Angelo, Scott Bergen, and Patricia Delamônica. 2001. Development of the Brazilian Amazon. *Science* 292(5522): 1651–1654.

Skidmore, Thomas E. 1999. *Brazil: Five Centuries of Change.* New York: Oxford University Press.

Smil, Vaclav. 2003. *Energy at the Crossroads: Global Perspectives and Uncertainties.* Cambridge, MA: MIT Press.

Snow, David A., and Robert D. Benford. 1992. Master Frames and Cycles of Protest. In *Frontiers in Social Movement Theory,* A. D. Morris and C. M. Mueller, eds. New Haven, CT: Yale University Press.

Soffer, Oren. 2004. Antisemitism, Statistics, and the Scientization of Hebrew Political Discourse: The Case Study of Ha-tsefirah. *Jewish Social Studies* 10.2(2004): 55–79.

Soto, Ana M., Honorato Justicia, Jonathan W. Wray, and Carlos Sonnenschein. 1991. *p*-Nonyl-Phenol: An Estrogenic Xenobiotic Released from "Modified" Polystyrene. *Environmental Health Perspectives* 92:167–173.

Sporn, M. B. 1996. The War on Cancer. *Lancet* 348(9025): 474.

Star, Susan Leigh, and James R. Griesemer. 1989. Institutional Ecology, "Translations" and Boundary Objects: Amateurs and Professionals in Berkeley's Museum of Vertebrate Zoology, 1907–39. *Social Studies of Science* 19:387–420.

Steinberg, Marc. 1998. Tilting the Frame: Considerations of Collective Action Framing from a Discursive Turn. *Theory and Society* 27(6): 845–872.

Stellman, S. D., and Q. S. Wang. 1994. Cancer Mortality in Chinese Immigrants to New York City: Comparison with Chinese in Tianjin and with White Americans. *Cancer* 73(4): 1270–1275.

Stilgoe, Jack. 2007. The (Co-)production of Public Uncertainty: UK Scientific Advice on Mobile Phone Health Risks. *Public Understanding of Science* 16(1): 45–61.

Summit Report. 2003. International Summit on Breast Cancer and the Environment: Research Needs, May 22–25, 2002. Chaminade, Santa Cruz, California.

Susan G. Komen Foundation. 2003. *Pathways to a Promise: Annual Report.*

Swirsky, Joan. 2005. *Map of Destiny: Pinpointing a Cancer Epidemic on the Kitchen Table.* Ashland, OH:Atlas Books.

Switkes, Glenn. 2002. Brazilian Government Pushes Ahead with Huge Dam in Amazon. *World Rivers Review* 17(3): 12–13.

———. 2007. *Brazilian Government Moves to Dam Principal Amazon Tributary.* Americas Program Report. Silver City, NM: International Relations Center.

Szklo, Alexandre Salem, Roberto Schaeffer, Marcio Edgar Schuller, and William Chandler. 2003. Brazilian Energy Policies Side-Effects on CO_2 Emissions Reduction. *Energy Policy* 33(3): 349–364.

Taub, Richard P. 1983. Urban Voluntary Associations: Locality Based and Externally Induced. *American Journal of Sociology* 83(2):425–442.

Taylor, Verta. 1989. Social Movement Continuity: The Women's Movement in Abeyance. *American Journal of Sociology* 54:761–775.

Tesh, Sylvia. 1988. *Hidden Arguments: Political Ideology and Disease Prevention Policy.* New Brunswick, NJ: Rutgers University Press.

———. 2000. *Uncertain Hazards: Environmental Activists and Scientific Proof.* Ithaca, NY: Cornell University Press.

Thiengo, S. C., S. B. Santos, and M. A. Fernandez. 2005. Freshwater Molluscs of the Lake of Serra da Mesa Dam, Goias, Brazil: I. Qualitative study. *Revista brasileira de zoologia* 22(4): 867–874.

Thomas, C. 2001. Habitat Conservation Planning: Certainly Empowered, Somewhat Deliberative, Questionably Democratic. *Politics Society* 29(1): 105–130.

Thompson, H. J. 1992. Effect of Amount and Type of Exercise on Experimentally Induced Breast Cancer. *Advances in Experimental Medicine and Biology* 322:61–71.

Treece, David. 2000. *Exiles, Allies, Rebels: Brazil's Indianist Movement, Indigenist Politics, and the Imperial Nation-State.* Boulder, CO: Westview Press.

Turaga, Uday. 2000. Damming Waters and Wisdom: Protest in the Narmada River Valley. *Technology in Society* 22:237–253.

Turner, Terence S. 1992. Defiant Images: The Kayapo Appropriation of Video. *Anthropology Today* 8(6): 5–16.

United Nations. Department of Economic and Social Affairs. 2001. *World Urbanization Prospects.* New York: United Nations Publications.

United States Postal Service. 2002. U.S. Postal Service to Revalue Breast Cancer Research Semipostal. Stamp Release #02-013.

Van der Windt, Henny J. and J.A.A. Swart. 2008. Ecological Corridors, Connecting Science and Politics: The Case of the Green River in the Netherlands. *Journal of Applied Ecology* 45(1): 124–132.

van Djick, Pitou, and Simon den Haak. 2007. *Troublesome Construction: IIRSA and Public-Private Partnerships in Road Infrastructure.* Berkeley: Centre for Latin American Studies and Documentation.

Vieira, Ubiratan Garcia. 2000. Limites do poder communicativo da argumentacao tecnica no licenciamento ambiental de hidreletricas em Minas Gerais. Master's Thesis, Universidade de Vicosa, Minas Gerais, Brazil.

Viola, Eduardo. 1997. The Environmental Movement in Brazil: Institutionalization, Sustainable Development, and the Crisis of Governance since 1987. In *Latin American Environmental Policy in International Perspective,* Gordon MacDonald, Daniel L. Nielson, and Marc A. Stern, eds. Boulder, CO: Westview Press.

Vogel, J. M. 2004. Tunnel Vision: The Regulation of Endocrine Disruptors. *Policy Sciences* 37(3–4): 277–303.

Voss, Kim. 1996. The Collapse of a Social Movement: The Interplay of Mobilizing Structures, Framing, and Political Opportunity in the Knights of Labor. In *Comparative Perspectives on Social Movements,* Doug McAdam, John D. McCarthy, and Mayer N. Zald, eds. Cambridge: Cambridge University Press.

Wade, Derick T., and Peter W. Halligan. 2004. Do Biomedical Models of Illness Make for Good Healthcare Systems? *British Medical journal* 329:1398–1401.

Warren, Mark. 1996. Deliberative Democracy and Authority. *American Political Science Review* 90(1): 46–60.

Weber, Max. 1952. *The Protestant Ethic and the Spirit of Capitalism.* Translated by Talcott Parsons. New York: Charles Scribner's Sons.

Weisman, Carol S. 1998. *Women's Health Care: Activist Traditions and Institutional Change.* Baltimore: Johns Hopkins University Press.

Welsh, Ian, Alexandra Plows, and Robert Evans. 2007. Human Rights and Genomics: Science, Genomics and Social Movements at the 2004 London Social Forum. *New Genetics and Society* 26(2): 123–135.

West, Dee W., Sally Glaser, and Angela Witt Prehn. 1998. *Status of Breast Cancer Research in the San Francisco Bay Area.* Union City, CA. Northern California Cancer Center.

Whittier, Nancy. 2002. Meaning and Structure in Social Movements. In *Social Movements: Identity, Culture and the State,* David S. Meyer, Nancy Whittier, and Belinda Robnett, eds. Oxford: Oxford University Press.

Windsor, J. E., and J. A. McVey. 2005. Nation within the Context of Large-Scale Environmental

Projects. *Geographical Journal* 171(2): 146–165.

Winner, Langdon. 1986. Do Artifacts Have Politics? In *The Reactor and the Whale: A Search for Limits in an Age of High Technology,* 19–39. Chicago: University of Chicago Press.

Wolff, Mary S., Paolo G. Toniolo, Eric W. Lee, Marilyn Rivera, and Neil Dublin. 1993. Blood Levels of Organochlorine Residues and Risk of Breast Cancer. *Journal of the National Cancer Institute* 85:648–652.

Wolford, Wendy. 2003. Producing Community: The MST and Land Reform Settlements in Brazil. *Journal of Agrarian Change* 3(4): 500.

Wolfson, Mark A. 2001. *The Fight against Big Tobacco: The Movement, the State, and the Public's Health*. New York: Aldine de Gruyter.

Woodhouse, E. J., and Dean Nieusma. 1999. Democratic Expertise: Integrating Knowledge, Power, and Participation. In *Policy Studies Annual,* R. Hoppe et al., eds.

World Bank. 1999. World Development Report. 1999. *Entering the 21st Century*. Washington, DC: World Bank.

World Commission on Dams. 2000. Dams and Development: A New Framework for Decision-Making. Washington, D.C.: World Bank.

Wrensch, Margaret, Terri Chew, Georgianna Farren, Janice Barlow, Flavia Belli, Christina Clarke, Christine A. Erdmann, Marion Lee, Michelle Moghadassi, Roni Peskin-Mentzer, Charles P. Quesenberry, Jr., Virginia Souders-Mason, Linda Spence, Marisa Suzuki, and Mary Gould. 2003. Risk Factors for Breast Cancer in a Population with High Incidence Rates. *Breast Cancer Research* 5(4): R88–R102.

Wright, Angus, and Wendy Wolford. 2003. *To Inherit the Earth: The Landless Movement and the Struggle for a New Brazil.* Oakland, CA: Food First Books.

Yach, Derek, and Stella Aguinaga Bialous. 2001. Junking Science to Promote Tobacco. *American Journal of Public Health* 91(11):1745–1748.

Zachariah, Matthew, and R. Sooryamoorthy. 1994. Science in Participatory Development: The Achievements and Dilemmas of a Development Movement; The Case of Kerala. London: Zed Books.

Zavestoski, Steve, Sabrina McCormick, and Phil Brown. 2005. Gender, Embodiment and Disease: Environmental Breast Cancer Activists' Challenge to Science, the Biomedical Model, and Policy. *Science as Culture* 13(4): 563–586.

Zhouri, Andréa. 2002. *Parecer sobre as informacoes compmentares ao EIA/ RIMA da UHE Murta.* Grupo de Estudos em Tematicas Ambientais. Belo Horizonte:Minas Gerais.

———. 2003. Hydroelectric Dams and Sustainability: Perspectives on the Use of Hydroelectric Energy in Brazil; Can Small Dams Help to Avoid Social and Environmental Problems? Paper presented to the Teuto-Brazilian Seminar on Renewable Energy, Berlin, June 2–3.

Zhouri, Andréa, Klemens Laschekski, and Doralice Barros Pereira, eds. 2005. *A insustentável leveza da política ambiental—Desenvolvimento e conflitos socioambientais*. São Paulo, Brazil: Autêntica.

Zhouri, Andréa, and Franklin Rothman. 2008. Assessoria aos atingidos por barragens em Minas Gerais. In *Vidas alagadas: Conflitos socioambientais, licenciamento e barragens,* Franklin Daniel Rothman, ed., 122–168. Viçosa: Editora UFV.

Index

Page numbers in *Italics* refer to figures and tables